Abstracts of The First Sourcebook on Asian Research in Mathematics Education: China, Korea, Singapore, Japan, Malaysia, and India

Special Supplement to the
International Sourcebooks in Mathematics and Science Education

Series Editor:
Bharath Sriraman, *The University of Montana*

International Sourcebooks in Mathematics and Science Education

Bharath Sriraman, Series Editor

The First Sourcebook on Nordic Research in Mathematics Education: Norway, Sweden, Iceland, Denmark, and Contributions From Finland (2010)
edited by Bharath Sriraman, Christer Bergsten, Simon Goodchild, Gudbjorg Palsdottir, Bettina Dahl Søndergaard, and Lenni Haapasalo

The First Sourcebook on Asian Research in Mathematics Education: China, Korea, Singapore, Japan, Malaysia, and India (2012)
Edited by Bharath Sriraman, Jinfa Cai, Kyeong-Hwa Lee, Fan Lianghuo, Yoshinori Shimizu, Lim Chap Sam, and K. Subramaniam

The First Sourcebook on Mediterranean and South Central Asian Research in Mathematics and Science Education: Cyprus, Israel, Turkey, Greece, Iran, Pakistan, and Central Asia (in development 2012)
edited by Bharath Sriraman, Constantinos Christou, Roza Leikin, Ahmet Arikan, Constantinos Tzakanis, and Anjum Halai

Abstracts of The First Sourcebook on Asian Research in Mathematics Education: China, Korea, Singapore, Japan, Malaysia, and India

Edited by

Bharath Sriraman
The University of Montana

Jinfa Cai
University of Delaware

Kyeong-Hwa Lee
Seoul National University

Fan Lianghuo
University of Southampton

Yoshinori Shimizu
University of Tsukuba

Lim Chap Sam
Universiti Sains Malaysia

K. Subramaniam
Tata Institute of Fundamental Research, India

Information Age Publishing, Inc.
Charlotte, North Carolina • www.infoagepub.com

ISBNS:

Paperback: 978-1-61735-825-8
Hardcover: 978-1-61735-826-5
eBook: 978-1-61735-827-2

Copyright © 2013 IAP–Information Age Publishing, Inc.

All rights reserved. No part of this publication may be reproduced, stored in a retrieval system, or transmitted in any form or by any electronic or mechanical means, or by photocopying, microfilming, recording or otherwise without written permission from the publisher.

Printed in the United States of America

Advisory Board

Ahmet Arikan, *Gazi University, Turkey*
Marcelo Borba, *São Paulo State University, Brazil*
Jinfa Cai, *University of Delaware*
Lim Chapsam, *Universiti Sains, Malaysia*
Lyn English, *Queensland University of Technology, Australia*
Viktor Freiman, *University of Moncton, Canada*
Simon Goodchild, *University of Agder, Norway*
Gudbjorg Palsdottir, *University of Iceland*
Guenter Toerner, *University of Duisburg-Essen Germany*
Doru Stefanescu, *University of Bucharest, Romania*
K. Subramaniam, *Tata Institute of Fundamental Research, Mumbai, India*

CONTENTS

CHINA

PART I: CULTURE, TRADITION, AND HISTORY

1. "Zhi Yì Xíng Nán (Knowing Is Easy and Doing Is Difficult)"
   or Vice Versa?: A Chinese Mathematician's Observation
   on History and Pedagogy of Mathematics Activities
   Man-Keung Siu 5

2. The Study on Application of Mathematics History
   in Mathematics Education in China
   Zezhong Yang and Jian Wang 7

3. Cultural Roots, Traditions, and Characteristics
   of Contemporary Mathematics Education in China
   Xuhui Li, Dianzhou Zhang and Shiqi Li 9

PART II: ASSESSMENT AND EVALUATION

4. Factors Affecting Mathematical Literacy Performance
   of 15-Year-Old Students in Macao: The PISA Perspective
   Kwok-Cheung Cheung 13

5. Has Curriculum Reform Made A Difference
   in the Classroom?: An Evaluation of the New Mathematics
   Curriculum in Mainland China
   Yujing Ni, Qiong Li, Jinfa Cai, and Kit-Tai Hau 15

6. Effect of Parental Involvement and Investment on Mathematics Learning: What Hong Kong Learned From PISA
   Esther Sui Chu Ho 17

PART III: CURRICULUM

7. Early Algebra in Chinese Elementary Mathematics Textbooks: The Case of Inverse Operations
   Meixia Ding 21

8. The Development of Chinese Mathematics Textbooks for Primary and Secondary Schools Since the Twentieth Century
   Shi-hu Lv, Ting Chen, Aihui Peng, and Shangzhi Wang 23

9. Mathematics Curriculum and Teaching Materials in China from 1950–2000
   Jianyue Zhang, Wei Sun, and Arthur B. Powell 25

10. Chinese Mathematics Curriculum Reform in the Twenty-first Century: 2000-2010
    Jian Liu, Lidong Wang, Ye Sun, and Yiming Cao 27

11. Basic Education Mathematics Curriculum Reform in the Greater Chinese Region: Trends and Lessons Learned
    *Chi-Chung Lam, Ngai-Ying Wong, Rui Ding,
    Siu Pang Titus Li, and Yun-Peng Ma* 29

12. Characterizing Chinese Mathematics Curriculum: A Cross-National Comparative Perspective
    Larry E. Suter and Jinfa Cai 31

PART IV: MATHEMATICAL COGNITION

13. Promoting Young Children's Development of Logical-Math Thinking Through Addition, Subtraction, Multiplication, and Division in Operational Math
    Zi-Juan Cheng 35

14. Development of Mathematical Cognition in Preschool Children
    Qingfen Hu and Jing Zhang 37

15. Chinese Children's Understanding of Fraction Concepts
    Ziqiang Xin and Chunhui Liu 39

16. Teaching and Learning of Number Sense in Taiwan
    Der-Ching Yang 41

17. Contemporary Chinese Investigations of Cognitive Aspects
    of Mathematics Learning
    Ping Yu, Wenhua Yu, and Yingfang Fu 43

18. Chinese Mathematical Processing and Mathematical Brain
    Xinlin Zhou, Wei Wei, Chuansheng Chen, and Qi Dong 45

PART V: TEACHING AND TEACHER EDUCATION

19. Comparing Teachers' Knowledge on Multidigit Division
    Between the United States and China
    Shuhua An and Song A. An 49

20. Problem Solving in Chinese Mathematics Education:
    Research and Practice
    Jinfa Cai, Bikai Nie, and Lijun Ye 51

21. Developing a Coding System for Video Analysis
    of Classroom Interaction
    Yiming Cao, Chen He, and Liping Ding 53

22. Mathematical Discourse in Chinese Classrooms:
    An Insider's Perspective
    Ida Ah Chee Mok, Xinrong Yang, and Yan Zhu 55

23. Reviving Teacher Learning: Chinese Mathematics
    Teacher Professional Development in the Context
    of Educational Reform
    Lynn W. Paine, Yanping Fang, and Heng Jiang 57

24. The Status Quo and Prospect of Research
    on Mathematics Education for Ethnic Minorities in China
    *Hengjun Tang, Aihui Peng, Bifen Chen, Yu Bo,
    Yanping Huang, and Naiqing Song* 59

25. Chinese Elementary Teachers' Mathematics Knowledge
    for Teaching: Role of Subject Related Training,
    Mathematic Teaching Experience, and Current
    Curriculum Study in Shaping Its Quality
    Jian Wang 61

26. Why Always Greener on the Other Side?: The Complexity
    of Chinese and U.S. Mathematics Education
    Thomas E. Ricks 63

PART VI: TECHNOLOGY

27. A Chinese Software SSP for the Teaching and Learning of Mathematics: Theoretical and Practical Perspectives
    Chunlian Jiang, Jingzhong Zhang, and Xicheng Peng 67

28. E-Learning in Mathematics Education
    Siu Cheung Kong 69

KOREA

29. Korean Research in Mathematics Education
    Kyeong-Hwa Lee, Jennifer M. Suh, Rae Young Kim, and Bharath Sriraman 73

30. A Review of Philosophical Studies on Mathematics Education
    JinYoung Nam 77

31. Mathematics Curriculum
    Kyungmee Park 79

32. Mathematics Textbooks
    JeongSuk Pang 81

33. Using the History of Mathematics to Teach and Learn Mathematics
    Hyewon Chang 83

34. Perspectives on Reasoning Instruction in the Mathematics Education
    BoMi Shin 85

35. Mathematical Modeling
    Yeong Ok Chong 87

36. Gender and Mathematics
    Eun Jung Lee 89

37. Mathematics Assessment
    GwiSoo Na 91

38. Examining Key Issues in Research on Teacher Education
    Gooyeon Kim 93

39. Trends in the Research on Korean Teachers' Beliefs About Mathematics Education
    Dong-Hwan Lee 95

SINGAPORE

40. A Review of Mathematical Problem-Solving Research Involving Students in Singapore Mathematics Classrooms (2001 to 2011): What's Done and What More Can be Done
    Chan Chun Ming Eric 99

41. Research on Singapore Mathematics Curriculum and Textbooks: Searching for Reasons Behind Students' Outstanding Performance
    Yan Zhu and Lianghuo Fan 103

42. Teachers' Assessment Literacy and Student Learning in Singapore Mathematics Classrooms
    Kim Hong Koh 107

43. A Theoretical Framework for Understanding the Different Attention Resource Demands of Letter-Symbolic Versus Model Method
    Swee Fong Ng 111

44. A Multidimensional Approach to Understanding in Mathematics Among Grade 8 Students in Singapore
    Boey Kok Leong, Shaljan Areepattamannil, and Berinderjeet Kaur ... 115

MALAYSIA

45. Mathematics Education Research in Malaysia: An Overview
    Chap Sam Lim, Parmjit Singh, Liew Kee Kor, and Cheng Meng Chew 121

46. Research Studies in the Learning and Understanding of Mathematics: A Malaysian Context
    Parmjit Singh and Sian Hoon Teoh 123

47. Numeracy Studies in Malaysia
    Munirah Ghazali and Abdul Razak Othman 125

48. Malaysian Research in Geometry
    Cheng Meng Chew 127

49. Research in Mathematical Thinking in Malaysia: Some Issues and Suggestions
    Shafia Abdul Rahman 129

50. Studies About Values in Mathematics Teaching
and Learning in Malaysia
Sharifah Norul Akmar Syed Zamri and Mohd Uzi Dollah *131*

51. Transformation of School Mathematics Assessment
Tee Yong Hwa, Chap Sam Lim, and Ngee Kiong Lau *133*

52. Mathematics Incorporating Graphics Calculator
Technology in Malaysia
Liew Kee Kor *135*

53. Mathematics Teacher Professional Development
in Malaysia
Chin Mon Chiew, Chap Sam Lim, and Ui Hock Cheah *137*

JAPAN

54. Mathematics Education Research in Japan:
An Introduction
Yoshinori Shimizu *141*

55. A Historical Perspective on Mathematics Education
Research in Japan
Naomichi Makinae *143*

56. The Development of Mathematics Education
as a Research Field in Japan
Yasuhiro Sekiguchi *147*

57. Research on Proportional Reasoning in Japanese Context
Keiko Hino *149*

58. Japanese Student's Understanding of School Algebra
Toshiakira Fujii *153*

59. Proving as an Explorative Activity
in Mathematics Education
Mikio Miyazaki and Taro Fujita *157*

60. Developments in Research on Mathematical Problem
Solving in Japan
Kazuhiko Nunokawa *161*

61. Research on Teaching and Learning Mathematics
With Information and Communication Technology
Yasuyuki Iijima *165*

62. "Inner Teacher": The Role of Metacognition
    in Learning Mathematics and Its Implication
    to Improving Classroom Practice
    Keiichi Shigematsu *167*

63. Cross-Cultural Studies on Mathematics Classroom Practices
    Yoshinori Shimizu *171*

64. Systematic Support of Life-Long Professional
    Development for Teachers Through Lesson Study
    Akihiko Takahashi *175*

INDIA

65. Evolving Concerns Around Mathematics as a School
    Discipline: Curricular Vision, Classroom Practice
    and the National Curriculum Framework (2005)
    Farida Abdulla Khan *181*

66. Curriculum Development in Primary Mathematics:
    The School Mathematics Project
    Amitabha Mukherjee and Vijaya S. Varma *185*

67. Intervening for Number Sense in Primary Mathematics
    Usha Menon *191*

68. Some Ethical Concerns in Designing Upper Primary
    Mathematics Curriculum: A Report From the Field
    Jayasree Subramanian, Sunil Verma, and Mohd. Umar *199*

69. Students' Understanding of Algebra
    and Curriculum Reform
    Rakhi Banerjee *207*

70. Professional Development of In-Service Mathematics
    Teachers in India
    Ruchi S. Kumar, K. Subramaniam, and Shweta Naik *213*

71. Insights Into Students' Errors Based on Data
    From Large-Scale Assessments
    Aaloka Kanhere, Anupriya Gupta, and Maulik Shah *219*

72. Assessment of Mathematical Learning:
    Issues and Challenges
    Shailesh Shirali *227*

73. Technology and Mathematics Education:
    Issues and Challenges 233
    Jonaki B. Ghosh *233*

74. Mathematics Education in Precolonial
    and Colonial South India
    Senthil Babu D. *243*
75. Representations of Numbers in the Indian Mathematical
    Tradition of Combinatorial Problems
    Raja Sridharan and K. Subramaniam *249*

CHINA

SECTION EDITOR
Jinfa Cai
University of Delaware

Advisory Board

Stephen Hwang, *University of Delaware*
Tammy Garber, *University of Delaware*
Bikai Nie, *University of Delaware*

Editorial Board

Shuhua An, *California State University-Long Beach*
Jianshen Bao, *East China Normal University*
Yiming Cao, *Beijing Normal University*
Kwok-cheung Cheung, *University of Macau*
Zaiping Dai, *Zhejiang Educational Institute*
Meixia Ding, *University of Nebraska-Lincoln*
Rongjin Huang, *Middle Tennessee State University*
Chunlian Jiang, *University of Macau*
Siu Cheung Kong, *Hong Kong Institute of Education*
Eddie Chi Keung Leung, *Hong Kong Institute of Education*
Shuk-kwan S. Leung, *National Sun Yat-sen University*
Frederick K.S. Leung, *University of Hong Kong*
Jun Li, *East China Normal University*

Shiqi Li, *East China Normal University*
Xuhui Li, *California State University-Long Beach*
Yeping Li, *Texas A&M University*
Fou-lai Lin, *National Taiwan Normal University*
Yuan-Horng Lin, *National Taichung University*
Jian Liu, *National Center for School Curriculum and Textbook Development*
Chuanhan Lv, *Guizhou Normal University*
Shi-hu Lv, *North West China Normal University*
Yun-Peng Ma, *North East China Normal University*
Lynn W. Paine, *Michigan State University*
Man-Keung Siu, *University of Hong Kong*
Rongbao Tu, *Nanjing Normal University*
Zhonghe Wu, *National University*
Ziqiang Xin, *Central University of Finance and Economics*
Ngai-Ying Wong, *Chinese University of Hong Kong*
Gong Yang, *People's Education Press and Institute of Curriculum Studies*
Lijun Ye, *Hangzhou Normal University*
Ping Yu, *Nanjing Normal University*
Wenfong Zhu, *Beijing Normal University*
Ying Zhou, *Guanxi Normal University*
Yingbo Zhang, *Beijing Normal University*
Jianyue Zhang, *People's Education Press and Institute of Curriculum Studies*

PART I

CULTURE, TRADITION, AND HISTORY

CHAPTER 1

"ZHI YÌ XÍNG NÁN (KNOWING IS EASY AND DOING IS DIFFICULT)" OR VICE VERSA?

A Chinese Mathematician's Observation on History and Pedagogy of Mathematics Activities

Man-Keung Siu
University of Hong Kong

ABSTRACT

In this essay the author shares with the readers, mainly through knowledge gleaned from his own participation in HPM (History and Pedagogy of Mathematics) activities, which he has experienced since the mid 1970s, with additional emphasis on the happenings in the Chinese community that he works in as a teacher at the tertiary level.

Keywords: mathematical history, pedagogy

CHAPTER 2

THE STUDY ON APPLICATION OF MATHEMATICS HISTORY IN MATHEMATICS EDUCATION IN CHINA

Zezhong Yang and Jian Wang
Shandong Normal University

ABSTRACT

This paper focuses on studies about the application of mathematics history in mathematics education in China, mainly introduces the studies about the application of mathematics history in mathematics textbooks and in mathematics teaching, and explicates the results of them. Meanwhile, this paper appraises these studies and puts forward some suggestions for further research.

Keywords: Mathematics history, education, mathematics textbooks, mathematics teaching, appli

CHAPTER 3

CULTURAL ROOTS, TRADITIONS, AND CHARACTERISTICS OF CONTEMPORARY MATHEMATICS EDUCATION IN CHINA

Xuhui Li
California State University-Long Beach

Dianzhou Zhang and Shiqi Li
East China Normal University

ABSTRACT

In this chapter, we first examine the overarching historical contexts of mathematics education in China and identify several of its most crucial cultural roots from economic, sociopolitical, administrative, and academic perspectives. Then we specifically discuss the influences of Confucian educational philosophies and traditions as well as the key features of traditional Chinese mathematics. After unfolding the modern development of mathematics education in China in the past century, we summarize the major characteristics of contemporary mathematics education in China and highlight their connections to the cultural factors and traditions analyzed earlier.

Keywords: mathematics education in China, history, culture, tradition, characteristics

PART II

ASSESSMENT AND EVALUATION

CHAPTER 4

FACTORS AFFECTING MATHEMATICAL LITERACY PERFORMANCE OF 15-YEAR-OLD STUDENTS IN MACAO

The PISA Perspective

Kwok-Cheung Cheung
Faculty of Education, University of Macau

ABSTRACT

Macao, a special administrative region of the People's Republic of China, participated in the PISA 2003 Mathematical Literacy Study. This seminal study, organized by OECD and involving 43 countries/economies, provides baseline information for all stake-holders in the educational arena to judge existing schooling practices and plan future mathematics education at both the primary and secondary education levels. This paper seeks to examine pertinent student and school factors affecting mathematical literacy performance in Macao secondary schools, after taking into account ESCS (i.e. Economic, Social and Cultural Status of the home), immigrant status, and gender of student. Issues pertaining to mathematics education in Macao schools are examined.

Keywords: mathematical literacy, ESCS, mathematics education, school effectiveness, PISA, Macao

CHAPTER 5

HAS CURRICULUM REFORM MADE A DIFFERENCE IN THE CLASSROOM?

An Evaluation of the New Mathematics Curriculum in Mainland China

Yujing Ni
The Chinese University of Hong Kong

Qiong Li
Beijing Normal University

Jinfa Cai
University of Delaware

Kit-Tai Hau
The Chinese University of Hong Kong

ABSTRACT

This chapter provides an account of a study which evaluated the effectiveness of China's new mathematics curriculum on teaching and learning in primary mathematics classrooms. It first provides a brief sketch of the main features of the new mathematics curriculum. It then explains the conceptualization guiding the evaluation study. Next, main findings of the evaluation are presented. The study provides much-needed, multi-facets, and longitudinal evidence about effects of the curriculum based on systematic classroom observations and multiple measures of student learning outcomes. It offers an example to illustrate a conceptual framework and research design for studying effects of curriculum reforms in mainland China.

Keywords: Curriculum reform, primary mathematics, curriculum evaluation, mathematics classroom instruction, student mathematics achievement

CHAPTER 6

EFFECT OF PARENTAL INVOLVEMENT AND INVESTMENT ON MATHEMATICS LEARNING

What Hong Kong Learned From PISA

Esther Sui Chu Ho
The Chinese University of Hong Kong

ABSTRACT

This chapter examines the effect of parent involvement and investment on students' mathematical literacy. The study identifies five types of involvement, namely, cultural communication, social communication, parenting, school communication and school volunteering, and three forms of parental investment, namely, cultural resources, educational resources, and computer facilities. The findings of this study suggest that certain forms of involvement and all three forms of investment are strong predictors of the two measures of students' mathematical literacy—mathematics performance and self-efficacy in mathematics—even after gender, immigrant status, family structure, and socio-economic status of students are taken into account.

The results indicate that parents as "motivators" through social and cultural communication with their children and parents as "resource providers" of all three forms of resources are significant contributors to students' mathematical literacy.

Keywords: parental involvement, parental investment, parent roles, mathematics performance, self-efficacy in mathematics, mathematical literacy, hierarchical linear modeling

PART III

CURRICULUM

CHAPTER 7

EARLY ALGEBRA IN CHINESE ELEMENTARY MATHEMATICS TEXTBOOKS

The case of Inverse Operations

Meixia Ding
University of Nebraska-Lincoln

ABSTRACT

This study examines one Chinese elementary textbook series' approach to the inverse relationships between addition and subtraction and between multiplication and division. A total of 342 instances are coded. Various types of problems under either computation or non-computation contexts are identified. All worked examples (designed for new learning) are situated in concrete situations to help students make sense of inverse relationships. The concreteness is then naturally connected to abstract representations. Finally, the Chinese textbooks space the learning of inverse operations over time. After a focused study in either first or second grade, both types of inverse relationships are subtly revisited across grades.

Keywords: early algebra, Chinese textbook approaches, inverse operations, addition and subtraction, multiplication and division

CHAPTER 8

THE DEVELOPMENT OF CHINESE MATHEMATICS TEXTBOOKS FOR PRIMARY AND SECONDARY SCHOOLS SINCE THE TWENTIETH CENTURY

Shi-hu Lv
Northwest Normal University

Ting Chen
Lanzhou City College

Aihui Peng
Institute of Higher Education, Southwest University

Shangzhi Wang
Institute of Mathematical Sciences, Capital Normal University

ABSTRACT

The reform of mathematics textbooks is closely related to that of politics, economics, culture, education, and similar factions. Described in chronological order, this paper categorizes the one hundred-year history of mathematic textbooks for primary and secondary schools into ten stages in terms of memorable historic events. At each stage, the background and the situation as well as the characteristics of the textbooks are analyzed.

Keywords: since the 20th century, China, mathematics textbooks for secondary schools, mathematics textbooks for primary schools, development

CHAPTER 9

MATHEMATICS CURRICULUM AND TEACHING MATERIALS IN CHINA FROM 1950–2000

Jianyue Zhang
People's Education Press, China

Wei Sun
Towson University, USA

Arthur B. Powell
Rutgers University, USA

ABSTRACT

Mathematics teaching materials has gone through many changes since the founding of the People's Republic of China. This chapter discusses issues related to the development of mathematics curriculum and mathematics teaching materials between 1950 and 2000, including the role of textbooks in mathematics education, guiding principles in content selection, design of textbook structure, organization of the mathematics contents, compilation process, field testing, as well as the significant changes at different times during this period. Specific mathematics contents in the textbooks and different approaches used by the Chinese curriculum developers are also introduced to help readers have a better understanding of the content knowledge Chinese students study.

Keywords: history of Chinese curriculum, curriculum design principles, textbook structure, mathematics teaching materials

CHAPTER 10

CHINESE MATHEMATICS CURRICULUM REFORM IN THE TWENTY-FIRST CENTURY

2000-2010

Jian Liu
National Center for School Curriculum and Textbook Development

Lidong Wang
Beijing Normal University

Ye Sun
West Virginia University

Yiming Cao
Beijing Normal University

ABSTRACT

This article discusses the current mathematics curriculum reform in China (2000-2010) from the perspective of curriculum policy. We summarize the background of Chinese Mathematics Curriculum Reform, the development, implementation and features of both the Mathematics Curriculum Standards for Full-time Compulsory Education and the Mathematics Curriculum Standards for Secondary Education. We also include the research results on the mathematics curriculum reform, contentions about the current mathematics curriculum among Chinese scholars, and future directions of the mathematics curriculum reform in China.

Keywords: mathematics, curriculum, reform, policy, China

CHAPTER 11

BASIC EDUCATION MATHEMATICS CURRICULUM REFORM IN THE GREATER CHINESE REGION

Trends and Lessons Learned

Chi-Chung Lam
Hong Kong Institute of Education

Ngai-Ying Wong
The Chinese University of Hong Kong

Rui Ding
Northeast Normal University

Siu Pang Titus Li
*Association for Childhood Education International—
Hong Kong & Macau*

Yun-Peng Ma
Northeast Normal University

ABSTRACT

Since the 1990s, Hong Kong, Taiwan, the Chinese mainland, and Macao have all launched curriculum reform one after another. The major thrust of the mathematics curriculum reform in these places was to enhance students' higher-order thinking skills and twenty-first century skills. Constructivist teaching and learning was adopted by the mathematics reformers as a means to achieve the reform goals. Yet the reform encountered different levels of resistance, which is a result of the interplay of the following factors in the four places: sharing of similar cultural beliefs in education, their special socio-political contexts, the history and development of education, and the mathematics curriculum. The winding road of mathematics curriculum change indicates the necessity of identifying the unique strength of mathematics teaching and abandoning the rationalist curriculum reform approach in the Chinese cultural region.

Keywords: mathematics curriculum reform, curriculum implementation, greater Chinese region, mathematics education

CHAPTER 12

CHARACTERIZING CHINESE MATHEMATICS CURRICULUM

A Cross-National Comparative Perspective

Larry E. Suter
National Science Foundation

Jinfa Cai
University of Delaware

ABSTRACT

This re-analysis of objective information about the timing and content of the mathematics curriculum for elementary and secondary students in China, Hong Kong, and the United States is propelled by previous survey results that found that mathematics achievement for students in China far exceeds that of other countries. It seeks evidence that the character of the curriculum might explain higher achievement in China. Data that had been collected about the intended and implemented curriculum in mathematics in 1995 for the TIMSS are analyzed to provide a deep description of the age that new mathematical content are introduced to elementary and secondary students. The TIMSS curriculum data set provides an expert description of the content areas intended to be taught at each age and an analysis of the content of textbooks in those grades. All three countries cover similar mathematics topics but at different ages. The U.S. curriculum appears to include early emphasis on more topics, such as equations and formulas, while the Chinese mathematics curriculum is more focused on fewer years of study for each topic than in the U.S. or Hong Kong. Newly introduced topics are not as frequently repeated as they are in the U.S. The mathematics curriculum in Hong Kong combined some characteristics of the curriculum in China with the U.S.

***Keywords*:** China, Hong Kong, curriculum, elementary school, framework, intended curriculum, international comparisons, TIMSS

PART IV

MATHEMATICAL COGNITION

CHAPTER 13

PROMOTING YOUNG CHILDREN'S DEVELOPMENT OF LOGICAL-MATH THINKING THROUGH ADDITION, SUBTRACTION, MULTIPLICATION, AND DIVISION IN OPERATIONAL MATH

Zi-Juan Cheng
The Chinese University of Hong Kong

ABSTRACT

In this chapter we introduce addition, subtraction, multiplication and division in *Operational Math for Preschool (OMP)* which is a Chinese culture based, teacher scaffolding, and aid-operation meditating program for 3- to 6-year-olds. The program uses several mediators, such as drawing picture to construct a series of images for representing math concepts and developing children's logical-math knowledge of part-part-whole relations of number in addition, subtraction, multiplication, and division and reverse relations between addition/multiplication and subtraction/division. The results of long-term research show the trajectories to promote young children to learn "big math," respond to the issues of what math in preschool should be, and how teaching affect young children's development of logical-math.

Keywords: preschoolers, part-part-whole relations of number, addition, subtraction, multiplication, division

CHAPTER 14

DEVELOPMENT OF MATHEMATICAL COGNITION IN PRESCHOOL CHILDREN

Qingfen Hu and Jing Zhang
*Institute of Developmental Psychology,
Beijing Normal University*

ABSTRACT

This chapter reviews the volume of developmental studies, conducted during the last three decades in China, that have focused on the mathematical cognition of preschool children. Among these studies, there are both investigational tests and experiments. Various research methods and approaches have been adopted to study the many facets of young children's mathematical cognition, including their concept of numbers, arithmetic skills, basic mathematical concepts, problem-solving strategies, as well as influencing factors in mathematical cognition.

Keywords: number concepts, calculation, basic mathematical concepts

CHAPTER 15

CHINESE CHILDREN'S UNDERSTANDING OF FRACTION CONCEPTS

Ziqiang Xin
Central University of Finance and Economics, Beijing, China

Chunhui Liu
Beijing Normal University, Beijing, China

ABSTRACT

Fractions are usually regarded as a difficult topic in school-mathematics. Thus, in the present chapter, we focus on how Chinese students understand the concept of fractions. Concretely, we address the following issues: (1) Chinese children's understanding of fractions in preschool, primary, and middle school; (2) their typical difficulties in understanding the fraction concept and the sources of the difficulties; (3) researchers' current controversies over fraction representation; and (4) directions in future fraction instruction and research. These concepts are further investigated cross-culturally.

Keywords: fraction concept, representation, whole number bias, cross-cultural study, mathematical achievement

CHAPTER 16

TEACHING AND LEARNING OF NUMBER SENSE IN TAIWAN

Der-Ching Yang
National Chiayi University

ABSTRACT

The major purpose of this study was to introduce the teaching and learning of number sense in Taiwan of the Huaren Region during the past fifteen years. Therefore, this chapter includes four parts. Part 1 introduces the meaning of number sense and its importance to mathematics teaching and learning. Part 2 reports the web-based two-tiered test of number sense for elementary school children. Part 3 reports the implementation of teaching and learning number sense in Taiwan. Lastly, Part 4 makes conclusions, implications, and suggestions for future research.

Keywords: Huaren Region, mathematics, number sense, teaching and learning, written computations

CHAPTER 17

CONTEMPORARY CHINESE INVESTIGATIONS OF COGNITIVE ASPECTS OF MATHEMATICS LEARNING

Ping Yu
Nanjing Normal University

Wenhua Yu

Yingfang Fu

ABSTRACT

In the 1960s, some psychologists began to study primary school students' development of concepts, such as the number concept and the space concept. As time went on, more investigations focused on primary school students' development of mathematical thinking and psychological mechanism of solving mathematical problems. Since the 1980s, more and more researches paid attention to secondary school students' cognitive learning of mathematics, concentrating on the aspects of mathematical thinking, learning transfer, problem solving, meta-cognition, and CPFS structure, et cetera.

Keywords: mathematical cognition, learning transfer, problem solving, meta-cognition, CPFS structure

CHAPTER 18

CHINESE MATHEMATICAL PROCESSING AND MATHEMATICAL BRAIN

Xinlin Zhou and Wei Wei
Beijing Normal University

Chuansheng Chen
University of California, Irvine

Qi Dong
Beijing Normal University

ABSTRACT

This chapter first presents a summary of the studies on mathematical cognition and mathematical brain that have been conducted in the Huaren regions (e.g, Mainland China, Taiwan, Hong Kong, and Macao). These studies demonstrated a culture-specific mathematical processing and mathematical brain as well as some general mechanisms. Then an empirical investigation on Chinese undergraduates' simple arithmetic is fully described. The investigation reveals how Chinese arithmetic processing, including simple addition and multiplication, is affected by arithmetic skills. Finally, the implications of previous studies in the Huaren regions and future research on the mathematical function of the brain are discussed.

Keywords: mathematical brain, mathematical cognition, cognitive arithmetic, numerical processing, problem size effect, operand order effect

PART V

TEACHING AND TEACHER EDUCATION

CHAPTER 19

COMPARING TEACHERS' KNOWLEDGE ON MULTIDIGIT DIVISION BETWEEN THE UNITED STATES AND CHINA

Shuhua An
California State University, Long Beach

Song A. An
Texas A&M University, College Station

ABSTRACT

This study compares U.S. and Chinese teachers on their content, pedagogical, and pedagogical content knowledge of multidigit division. The findings show that Chinese teachers use estimation as a main approach to foster students' conceptual understanding and computation fluency in multidigit division. In this approach, they would integrate, review, and scaffold their instruction while also allowing the students to try the division first. On the other hand, U.S. teachers indicate that they use algorithms in multidigit division after they explain the concept. The study demonstrates an empirical way to quantitatively measure teachers' pedagogical content knowledge in a statistical analysis to determine a degree of the relationship between content and pedagogical knowledge. The results show that Chinese teachers, not U.S. teachers, have established pedagogical content knowledge in teaching multidigit division.

Keywords: teacher knowledge in mathematics, multidigit division, estimation, algorism, place value, review and scaffolding

CHAPTER 20

PROBLEM SOLVING IN CHINESE MATHEMATICS EDUCATION

Research and Practice

Jinfa Cai and Bikai Nie
University of Delaware

Lijun Ye
Hangzhou Normal University

ABSTRACT

This chapter attempts to paint a picture of problem solving in Chinese mathematics education, where problem solving has been viewed both as an instructional goal and as an instructional approach. In discussing problem-solving research from four perspectives, it is found that the research in China has been much more content and experience-based than cognitive and empirical-based. We also describe several problem-solving activities in the Chinese classroom, including "one problem multiple solutions," "multiple problems one solution," and "one problem multiple changes." Unfortunately, there are no empirical investigations that document the actual effectiveness and reasons for the effectiveness of those problem-solving activities. Nevertheless, these problem-solving activities should be useful references for helping students make sense of mathematics.

Keywords: problem solving, Chinese mathematics education, research, practice

CHAPTER 21

DEVELOPING A CODING SYSTEM FOR VIDEO ANALYSIS OF CLASSROOM INTERACTION

Yiming Cao
Beijing Normal University

Chen He
Beijing Huiwen Middle School

Liping Ding
Shanghai Soong Ching Ling Child Development Research Center

ABSTRACT

This chapter focuses on two of three Grade 7 (Year 8) mathematics classrooms in a larger study that was part of the national component of the China (Shanghai) Learner's Perspective Study (LPS). An attempt has been made to develop a Studiocode coding system to enable researchers to systematically analyze the pattern of classroom interactions in mathematics lessons. The initial findings of this chapter are concerned with the main types of the teacher and students interactions and the main forms of classroom interaction according to the teacher's and students' behaviors in the two classrooms.

Keywords: classroom interaction, Studiocode, mathematics, lower secondary school, China

CHAPTER 22

MATHEMATICAL DISCOURSE IN CHINESE CLASSROOMS

An Insider's Perspective

Ida Ah Chee Mok
The University of Hong Kong

Xinrong Yang and Yan Zhu
East China Normal University

ABSTRACT

We attempt to investigate what mathematical discourse in Chinese classrooms may be like. Firstly, we seek a definition for mathematical discourse based on literature. A differentiation between discourse and mathematical discourse will be made. Secondly, we present a summary of studies on mathematical discourse in Chinese classrooms. The literature review will be structured around the following (1) Chinese mathematics lesson structure, (2) patterns of interactions in Chinese mathematics classrooms, and (3) instructional styles. The emerging features include I-R-F, funnel and focusing patterns, teaching-leading, coherence, and mathematical formality.

Keywords: mathematical discourse, Chinese mathematics classrooms, interaction, instructional style, coherence, mathematical language

CHAPTER 23

REVIVING TEACHER LEARNING

Chinese Mathematics Teacher Professional Development in the Context of Educational Reform

Lynn W. Paine
Michigan State University

Yanping Fang
National Institute of Education, Singapore

Heng Jiang
Michigan State University

ABSTRACT

In this paper, we examine the ways in which recent developments for Chinese mathematics teachers involve building on common forms of teacher learning to reinvigorate practices, enliven professional development, and support a kind of teaching congruent with the demands of the new mathematics curriculum. We argue that both the top-down and bottom-up forces have critiqued some of the traditional practices of teacher development and advocated for new efforts—some of which reveal the influence of a global circulation of ideas—that create the possibilities for reviving teacher learning in mathematics. We consider how Action Education (*xing dong jiao yu*) was introduced and has supported the coupling of school-level teacher research with teacher learning in mathematics. This case adds to our understanding of what effective mathematics teacher learning may entail and helps raise important questions regarding the impact of local context and cultural practices on teacher learning.

Keywords: mathematics teacher professional development; *xing dong jiao yu*

CHAPTER 24

THE STATUS QUO AND PROSPECT OF RESEARCH ON MATHEMATICS EDUCATION FOR ETHNIC MINORITIES IN CHINA

Hengjun Tang
Zhejiang Normal University, China

Aihui Peng
Southwest University, China

Bifen Chen
Zhejiang Normal University, China

Yu Bo, Yanping Huang, and Naiqing Song
Southwest University, China

ABSTRACT

Mathematics education for ethnic minorities is an important topic in mathematics education, which is particularly essential for China, a multinational country consisting of 56 nationalities. Over the past 30 years, numbers of research studies have been conducted on this topic in China. This chapter introduces the current research status of mathematics education for ethnic minorities in China by reviewing existing research findings. It includes both a systemic review targeted to answer what has been done and what problems there are in this field. This paper also specifically reviews the following four issues: mathematics textbooks and curricula for ethnic minorities, minority students' mathematics learning, professional development of minority mathematics teachers, and the integration of ethno-mathematics culture into mathematics curricula and teaching. Based on the review, further research directions and tendencies are also discussed.

Keywords: mathematics education for ethnic minorities in China, research status quo and prospect, bilingual education, mathematics learning of minority students, professional development of ethnic mathematics teachers, ethno-mathematics culture

CHAPTER 25

CHINESE ELEMENTARY TEACHERS' MATHEMATICS KNOWLEDGE FOR TEACHING

Role of Subject Related Training, Mathematic Teaching Experience, and Current Curriculum Study in Shaping Its Quality

Jian Wang
University of Nevada, Las Vegas

ABSTRACT

Sound mathematics knowledge for teaching is seen as crucial for teachers' effective mathematics instruction and thus for improving student mathematics learning. There are different assumptions about the role of mathematics and mathematics education, mathematics teaching experiences, and curriculum for teaching in shaping teachers' mathematics knowledge for teaching. Chinese elementary teachers are found to have deep mathematics knowledge for teaching. Their various professional backgrounds and curriculum context provide a unique situation to examine these assumptions. This study analyzes the survey and curriculum data from Chinese elementary mathematics teachers and finds that the participants' mathematics and mathematics education training and mathematics teaching experience alone have limited influences on the quality of their mathematics knowledge for teaching. Their reconstruction of mathematics understanding of mathematical concepts that they have to teach under the centralized curriculum and in collaboration with their colleagues importantly shaped the quality of their mathematics knowledge for teaching.

Keywords: mathematics education, teacher knowledge, and Chinese teachers

CHAPTER 26

WHY ALWAYS GREENER ON THE OTHER SIDE?

The Complexity of Chinese and U.S. Mathematics Education

Thomas E. Ricks
Louisiana State University

ABSTRACT

Researchers have noted that both China and the United States seem to be striving to copy aspects of the other's educational system. In an attempt to understand this phenomenon, the complexities of mainland China's educational factors were juxtaposed against those of the United States. Analysis revealed that China's complexity currently occurs mostly at the *supra-individual* level (familial, class cohorts, coherent lessons, professional development cooperatives, etc.), while the United State's complexity is manifest primarily by enabling the *individual* student. China envies U.S. students' creativity, independence, curiosity, and innovation while the United States seeks to adopt the integrated features of China's educational system.

Keywords: China, complexity, cross-culture comparison, mathematics education, reform, United States

PART VI

TECHNOLOGY

CHAPTER 27

A CHINESE SOFTWARE SSP FOR THE TEACHING AND LEARNING OF MATHEMATICS

Theoretical and Practical Perspectives

Chunlian Jiang
University of Macau, Macau

Jingzhong Zhang
Central China Normal University and Guangzhou University

Xicheng Peng
Central China Normal University

ABSTRACT

SSP is the first dynamic geometry software designed by Chinese scholars in 2004 for the teaching and learning of mathematics. Twenty-six schools were involved in an experimental project "Application of Z+Z Intelligent Education Platforms into the National Mathematics Education Reform," which started in 2003. Since then, more and more teachers have used it as a teaching tool, and more and more students have used it as a learning tool in the teaching and learning of mathematics. Starting from 2007, we have also introduced a course titled "Dynamic Geometry" to preservice teachers in normal universities in China. This paper will provide an overview of its design, characteristics, and practical applications in schools and in teacher education.

Keywords: SSP, Dynamic Geometry, IT in mathematics education

CHAPTER 28

E-LEARNING IN MATHEMATICS EDUCATION

Siu Cheung Kong
The Hong Kong Institute of Education

ABSTRACT

One of the aims of mathematics education is to empower learners to connect and explain relationships between mathematical symbols and their conceptual meaning. The capacities of digital technology for visual representation and graphical manipulation support learners to link procedural knowledge and conceptual understanding. This chapter discusses the opportunities of e-learning, which refers to the use of digital technology to facilitate learning for mathematics education from three perspectives, namely using computer-supported cognitive tools for promoting knowledge construction, using mobile technology for promoting collaborative learning, and using social networking technology for promoting extended learning.

Keywords: cognitive tool, collaborative learning, conceptual understanding, e-learning, mathematics education, mobile technology supported classroom, procedural knowledge, social networking technology

KOREA

SECTION EDITORS
Kyeong Hwa Lee
Seoul National University
&
Bharath Sriraman
The University of Montana

Editorial Board

Jennifer M. Suh, *George Mason University*
Rae Young Kim, *Ewha Womans University*
JinYoung Nam, *Gyeongin National University of Education*
Kyungmee Park, *Hongik University*
JeongSuk Pang, *Korea National University of Education*
BoMi Shin, *Chonnam National University*
Yeong Ok Chong, *Gyeongin National University of Education*
Hyewon Chang, *Chinju National University of Education*
GwiSoo Na, *Cheongju National University of Education*
Eun-Jung Lee, *Seoul National University*
Gooyeon Kim, *Sogang University*
Dong-Hwan Lee, *Korea Foundation for the Advancement of Science & Creativity*

CHAPTER 29

KOREAN RESEARCH IN MATHEMATICS EDUCATION

Kyeong-Hwa Lee
Seoul National University

Jennifer M. Suh
George Mason University

Rae Young Kim
Ewha Womans University

Bharath Sriraman
The University of Montana

SUMMATIVE ABSTRACT

On the basis of a principle "*Hong Ik In Gan*," which means to benefit all humankind, the goal of mathematics education of Korea is to nurture qualified persons to play lead roles in the future through providing personality and creativity education. Education in Korea, deeply rooted in Confucianism and Buddhism, is valued above everything else in the country. A teacher, whose literal meaning is "the one who was born ahead" in the Korean language, is considered to play a fundamental role in education by being an exemplary. In accordance with Korean sociocultural belief, which acknowledges the significance of education, mathematics education research society aims to build Korean perspectives, methodology, and strategies in mathematics education by fostering competent mathematics researchers, teachers, and learners

Korean students, like other East Asian students, are observed that they show a passive attitude when participating in mathematics classrooms (Leung & Park, 2002). However, they are highly motivated in reality and show the high achievement by continued effort. Mathematics teachers in Korea seem to have profound knowledge of mathematics content and pedagogy, and the educational zeal of Korean parents is also quite high. The Korean government has supported research to improve the quality of mathematics education with continued interest and attempted educational reform by the centralization of control (Cho, 2009; Lee, 2010). These socio-cultural backgrounds have played no small part in mathematics education research in Korea. Studies on mathematics curriculum, textbooks, and assessments are well researched, and themes related to mathematical reasoning, modeling, and the use of the history of mathematics are also studied with multifaceted perspectives. The government intends to improve the quality of teacher training and diversify the training programs and so various studies on teacher training and teacher's beliefs are also conducted. However, there is a lack of research on our country's own philosophy of mathematics education or gender-related issues. Studies on the aforementioned themes that have been done in Korea are reported in this book.

The chapter entitled *A review of philosophical studies on mathematics education* illustrates the philosophical backgrounds that have laid underneath mathematics education in Korea and the ways in which they have shaped and/or influenced the educational goals and blueprints of mathematics education in Korea. This chapter introduces both philosopher-centered work (Plato, Kant, Dewey, and Lakatos) and theme-centered work (constructivism, freedom, and humanity education) that have been committed to looking for the objectives and methods of mathematics education.

Shedding light on the national curriculum for K-12 school mathematics as an important factor to understand mathematics education in Korea, the chapter entitled *Mathematics curriculum* presents a brief overview of the Korean national curriculum for mathematics, including the history of the development of the mathematics curriculum, and the main features of the structure and content of the current curriculum (which was revised in 2007).

The chapter entitled *Mathematics textbooks* provides brief but concise information about Korean mathematics textbooks as well as research trends on the curriculum and mathematics textbooks in Korea by reviewing research papers published in seven major domestic journals in Korea during the past five years (2006-2010). This chapter describes the kinds of research that have been conducted with regard to Korean mathematics textbooks, i.e., analyses of Korean mathematics textbooks, international comparative studies on mathematics textbooks, and the examination of textbooks in other countries.

The chapter on using *The History of Mathematics to Teach and Learn Mathematics* takes an interesting look at how the history of mathematics is used to enrich the learning of mathematics. The paper focuses on the cognitive and affective purposes for using history in the process of learning mathematics. It also describes practical application of the history of mathematics in textbooks and research on the history of Korean mathematics and its use in mathematics education.

The chapter entitled *Perspectives on Reasoning Instruction in Mathematics Education* provides some examples of how reasoning is encouraged through "induction and analogy" and "justification and demonstration" on the discovery in the Korean mathematics curriculum. The author also suggests that more research is needed in the area of developing methods for teaching reasoning and strategies for reasoning instruction in the mathematics education in Korea.

The chapter called *Mathematical Modeling* begins with two definition of modeling (Lesh & Doerr, 2003 and Gravemeijer, 1997) and describes several examples of studies that focus on the meanings and stages of mathematical modeling. The author describes mathematical modelling as a way to seek reasonable explanations that deals with real situation and provides several examples of mathematical modeling activity such as the model in examining the changes in the number of hawks and pigeons and draw a graph by presenting a formula that models the predator-prey relations. The author describes mathematical modeling activity as a way to encourage students' mathematical thinking and understanding.

The chapter on *Gender and Mathematics* takes a closer look at curriculum tracks, course taking patterns, general aptitudes, and ratio in female participation in natural sciences and engineering. This paper also provides background into the social and cultural factors of Korea that might cause the underrepresentation of women in natural sciences and engineering careers. The mentoring community and intervention program called Women into Science and Engineering (WISE) is highlighted as an establishment of a network of women professions in STEM that encourages female students to pursue natural sciences and engineering careers.

The chapter entitled *Mathematics assessment* describes four different themes with regard to the assessments in Korea: the characteristics of Korean students' mathematical ability recognized in international assessments; the reality of the assessment of students' mathematics achievement; the current issues in mathematics assessment; and suggestions for future improvements in mathematics assessment. This paper provides useful information about mathematics assessments conducted in Korea at different levels (local, national, and international levels) from an insider's perspective.

Reviewing extensive empirical studies on teacher education in Korea, the chapter entitled *Examining key issues in research on teacher education* that only a small number of studies have been conducted around the topic of teacher education despite the importance of the topic in mathematics education. Popular topics in the literature on teacher education in Korea have included the mathematical content knowledge of teachers, teachers' beliefs and orientations towards mathematics and mathematics teaching, teachers' awareness of various issues around mathematics education, mathematics teaching efficacy, the professionalism of mathematics teachers, and teacher education programs for pre-service teachers.

The chapter entitled *Trends in the Research of Teachers' Beliefs about Mathematics Education* first identifies the characteristics of Korean teachers' beliefs about mathematics, its teaching and learning. Next it examines the sources that influence teachers' beliefs and how teachers construct their beliefs based on their experiences. Finally, the chapter explores the relationship between Korean teachers' beliefs and their teaching practices.

REFERENCES

Leung, F. K. S., & Park, K. M. (2002). Competent students, competent teachers? *International Journal of Educational Research, 37*(2), 113-129.

Lee, K. H. (2010). Searching for Korean perspective on mathematics education through discussion on mathematical modeling. *Journal of Educational Research in Mathematics, 20*(3), 221-239.

Cho, Y. D. (2009). Understanding Korean middle and secondary school classroom teaching and teachers' instructional activities: "Optimization behavior for teaching." *Proceedings of the 35th Conference of the Korea Society of Educational Studies in Mathematics* (pp. 1-27). Seoul, Korea.

CHAPTER 30

A REVIEW OF PHILOSOPHICAL STUDIES ON MATHEMATICS EDUCATION

JinYoung Nam
Gyeongin National University of Education

ABSTRACT

A theory of the objectives, content, and methods of mathematics education necessarily relies on a certain philosophy of mathematics education, and discussions on the theory certainly reflect the culture in which it is discussed. This paper reviews philosophical work on mathematics education in South Korea. Korean mathematics educators generally conduct their philosophical work as PhD theses, because of the depth and the width of the study. Thus, this paper reviews nine doctoral theses at universities in Korea. Some of the theses discuss the objectives, content, and methods of mathematics education through the perspective of a particular philosopher: Plato, Kant, Dewey, or Lakatos. Others deal with philosophical issues in mathematics education such as constructivism, freedom, or humanity education. Some works embrace oriental philosophies, Buddhism, or neo-Confucianist theory. This review is comprised of two parts: philosopher-centered work and theme-centered work.

Keywords: Philosophical perspectives, history of education, South Korea, Confucianism, Lakatos, Dewey

CHAPTER 31

MATHEMATICS CURRICULUM

Kyungmee Park
Hongik University

ABSTRACT

Korea, like many other Asian countries, has a uniform curriculum administered at the national level. To ensure the quality of education, the Education Law prescribes the curriculum for each school level and the criteria for the development of textbooks and instructional materials. Since mathematics textbooks are written strictly based on mathematics curriculum and these textbooks are the main resource of mathematics lessons, mathematics curriculum is one of the most important lenses through which to look into mathematics education in Korea. The mathematics curriculum in Korea has been revised periodically to reflect emerging needs of a changing society. This chapter begins with a brief overview of the Korean school mathematics curriculum, followed by an explanation of the details of the current curriculum which was revised in 2007. In the last section, some issues regarding future mathematics curriculum in Korea will be discussed.

Keywords: mathematics education, mathematics curriculum, law and education in South Korea, school mathematics, educational reform in Korea

CHAPTER 32

MATHEMATICS TEXTBOOKS

JeongSuk Pang
Korea National University of Education

ABSTRACT

Mathematics textbooks greatly influence what and how students learn, because almost all Korean teachers use them as their main instructional resources. Thus, the development of quality textbooks has been an important concern in shaping Korean mathematics instruction and, in fact, has prompted many studies on Korean mathematics textbooks. This chapter first provides an overview of Korean mathematics textbooks in terms of their development and characteristics. It then addresses recent research trends on mathematics textbooks and closes with key issues on the future development and research of mathematics textbooks.

Keywords: textbook development, textbook characteristics, research trend on mathematics textbooks, Korean mathematics textbooks

CHAPTER 33

USING THE HISTORY OF MATHEMATICS TO TEACH AND LEARN MATHEMATICS

Hyewon Chang
Chinju National University of Education

ABSTRACT

As mathematics is considered as not only logico-deductive output, but also as a plausible and inductive process, mathematics educators have begun to pay attention to the origin of mathematical knowledge, the necessity of the knowledge, and the process of invention, and to use them to teach and learn mathematics. The educational purposes of using history can be roughly classified into two categories: cognitive and affective. The former is to use their origins or developmental processes in teaching and learning some mathematical knowledge—mathematical terminology, algorithms, or concepts. The latter is to use history for motivating students, stimulating their curiosity, or encouraging them to have positive mathematical attitudes in the process of learning mathematics. Since the last part of the twentieth century, we have concentrated on the history of mathematics and its educational use both theoretically and practically in Korea. Much research has been conducted in respect to using history in mathematics education and its results have been applied to educational practice. This paper shows the practical application of using the history of mathematics in school mathematics, as well as theoretical studies in the history of mathematics and using its history in mathematics education.

Keywords: history of mathematics, history of Korean mathematics, Chosun mathematics books, history in mathematics education, historico-genetic principles, mathematics instruction

CHAPTER 34

PERSPECTIVES ON REASONING INSTRUCTION IN THE MATHEMATICS EDUCATION

BoMi Shin
Chonnam National University

ABSTRACT

Reasoning is the line of thought adopted to produce assertions and reach conclusions in task solving (Lithner, 2008). In addition, reasoning has many functions in mathematics including verification, explanation, systematization, discovery, communication, construction of theory, and exploration (Yackel & Hanna, 2003). Such mathematical reasoning is categorized as deductive and heuristic reasoning. Since ancient Greece times, deductive reasoning has been considered the best educational method for training minds. However, Polya (1954) pointed out that heuristic reasoning in addition to deductive reasoning should be seriously considered in mathematics education. According to Woo (2000), both deductive and heuristic reasoning should be addressed in the instruction of mathematical reasoning. That is, deductive reasoning and heuristic reasoning in the instruction of mathematical reasoning need to be dealt with in a balanced way. It is now widely accepted that mathematical knowledge spawns from inductive reasoning, analogy, and generalization, and is verified through deductive reasoning. This study analyzes the mathematics curriculum of Korea in terms of the deductive and heuristic reasoning derived from previous study results and introduces mathematics education learning/teaching trends in respect to reasoning.

Keywords: mathematical reasoning, heuristic reasoning, inductive versus deductive reasoning

CHAPTER 35

MATHEMATICAL MODELING

Yeong Ok Chong
Kyeongin National University of Education

ABSTRACT

Internationally, research on mathematical modeling seems to have originated from searching for alternatives after the failure of the New-Math approach toward mathematics education. But research on mathematical modeling in Korea began with an article by Ju (1991), Reflections on Teaching Modeling which established the foundations of mathematical modeling through a theoretical investigation in Korea. Thereafter, along with theoretical research, quantitative researches as well as qualitative research have continued on this topic. In this chapter, I will illuminate on the perspectives, meanings and stages of mathematical modeling discussed in Korea. Also I will present seminal research on this topic and tasks introduced in textbooks with respect to mathematical modeling that have been influenced by this research.

Keywords: mathematical modeling, stages of mathematical modeling, mathematization, model refinement

CHAPTER 36

GENDER AND MATHEMATICS

Eun Jung Lee
Seoul National University

ABSTRACT

Gender in mathematics education started to attract the interest of researchers in the 1970s due to a significant gender gap in favor of males in mathematics achievement. Studies on gender and mathematics have been actively conducted for the last over 30 years; yet, the studies have mainly been conducted in developed countries. This paper briefly discusses the gender differences in mathematics that play important role in studying science and engineering from two points of view: mathematics achievement and participation in mathematics and math-related fields. The social and cultural factors of Korea that might cause the underrepresentation of women in natural sciences and engineering careers are also discussed and finally, the intervention program: Women into Science and Engineering (WISE), carried out in Korea since 2002 to encourage female students to pursue natural sciences and engineering careers is presented along with its effectiveness.

Keywords: gender difference; mathematics achievement; male dominated societies; socio-cultural factors, WISE

CHAPTER 37

MATHEMATICS ASSESSMENT

GwiSoo Na
Cheongju National University of Education

ABSTRACT

In this chapter, various aspects of mathematics assessment in Korea are introduced. This chapter is consisted of four parts; the characteristics of Korean students' mathematical ability recognized in international assessments; the reality of assessment of students' mathematics achievement in Korea; the current issues in mathematics assessment in Korea; and future improvements in mathematics assessment in Korea. In the first part on the characteristics of Korean students' mathematical ability recognized in international assessments, the strengths and weaknesses of Korean students shown in the results of TIMSS and PISA are reported. In the second part on the reality of assessment of students' mathematics achievement in Korea, the National Assessment of Educational Achievement (NAEA) presently carried out on the national level and the assessment of student achievement administered on the school level are examined. In the third part on the current issues in mathematics assessment in Korea, the current trends of research in mathematics assessment and the current issues in mathematics assessment are discussed. In the last part on the future improvements in mathematics assessment in Korea, two emphases on the future improvements, that is, helping students to grow the mathematical literacy and to get the positive attitude toward mathematics are discussed.

Keywords: mathematics assessment, mathematical achievement, Korea, NAEA, PISA, TIMSS

CHAPTER 38

EXAMINING KEY ISSUES IN RESEARCH ON TEACHER EDUCATION

Gooyeon Kim
Sogang University

ABSTRACT

This section provides a review of research studies on teacher education in South Korea. Herein, teacher education includes professional development programs for inservice teachers as well as teacher education programs for preservice teachers. To gather empirical research studies on teacher education, I used the electronic data-base search engine, limiting the years of publication from 2000 to 2010. This search identified about 50 studies on the issue of teacher education. Although the body of literature might not be exhaustive, I only targeted empirical studies from the literature. In other words, I excluded conceptual papers discussing various conceptual ideas and framework as well as review papers on specific topics. Finally, I ended up analyzing about 40 empirical research studies. I examined the body of literature by reading and summarizing, ultimately analyzing it by focusing on one specific question: what are the specific issues that have been investigated in the field of mathematics education in South Korea for the past decade? In the following section, I present what I have found from the analysis and what implications can be gleaned from the review.

Keywords: mathematics teacher education, meta-review of professional development studies, empirical studies on teacher education, Korea

CHAPTER 39

TRENDS IN THE RESEARCH ON KOREAN TEACHERS' BELIEFS ABOUT MATHEMATICS EDUCATION

Dong-Hwan Lee
Korea Foundation for the Advancement of Science & Creativity

ABSTRACT

This paper offers an overview of recent researches on teachers' beliefs about mathematics education in Korea. The researches consist of three interrelated strands of works. The first strand identifies the characteristics of Korean teachers' beliefs about mathematics, its teaching and its learning from a variety of perspectives. The second strand, which is based on the first strand, investigates various potential sources of teachers' beliefs and attempts to understand the construction process of teachers' beliefs. The third strand, which can be seen as the ultimate goal of studying the first and second strands, explores the relationship between teachers' beliefs and their teaching.

Keywords: teachers' beliefs, Korea, mathematics teaching, teacher education

SINGAPORE

SECTION EDITOR
Lianghuo Fan
The University of Southampton, UK

Editorial Board

Lianghuo Fan, *The University of Southampton, UK*
Swee Fong Ng, *Nanyang Technological University, Singapore*
Lionel Pereira-mendoza, *Educational Consultant, Canada*

CHAPTER 40

A REVIEW OF MATHEMATICAL PROBLEM-SOLVING RESEARCH INVOLVING STUDENTS IN SINGAPORE MATHEMATICS CLASSROOMS (2001 TO 2011)

What's Done and What More Can be Done

Chan Chun Ming Eric
*National Institute of Education,
Nanyang Technological University, Singapore*

ABSTRACT

Since mathematical problem solving has been positioned to be the core of the Singapore Mathematics Curriculum Framework at the beginning of 1990, there has been heightened interest in research in the area of mathematical problem solving. Although several reviews of research in mathematics education have been carried out over different periods of time, this chapter synthesizes the review of research in mathematical problem solving involving students in the last decade (2001 to 2011).

This chapter begins by introducing a brief background to how the Singapore Mathematics Curriculum Framework (popularly known as the Singapore problem-solving framework) of today came about. As a centralized curriculum, it has adapted itself well with the changing mathematics education landscape from one that focuses on basic skills and knowledge towards embracing thinking skills and heuristics. Increasingly, it aims to develop students through more integrated approaches where the processes involved in problem solving—reasoning, communicating and making connections, are valued. The review of problem-solving research involving students based on several strands then follows.

The review begins with studies related to primary students' employment of the Model-Drawing Method. This method has been actively promoted as a tool to solve arithmetic and algebraic word problems since the 1980s. Research studies were carried out to understand the features of this tool that makes it suitable to solve word problems as well as the challenges students face when using it. As the Model-Drawing Method is also deemed to help students solve algebra word problems without the use of algebra, a research looks into whether students would prefer to use such a method when they are in secondary grade levels where symbolic algebra is taught while another research involves easing the transition between the pictorial representational aspect of the Model Drawing Method and the symbolic aspect through the use of a software. Besides the Model-Drawing Method, there are several research studies on students' use of Polya's heuristics. A research project by the Centre for Research in Pedagogy and Practice sought to develop the repertoire of heuristics for mathematical problem solving. Other research focused on the problem-solving ability and the impact of heuristics instructions in solving non-routine problems.

As the Ministry of Education continues to set future directions in the new millennium, a strand of research that surfaced includes students' solving of open-ended and real-world problems. In this respect, research interests have been broad-based and were found to focus on students' ability in open-ended problem solving, the enhancing of critical thinking, the relation to affect, the reasoning processes as well as students' model development. Though limited, these studies are a good start in itself as it presents a diversity of newer research that seeks to promote students' 21st century skills that teacher-practitioners must come to grips with.

Student success in problem solving is the pride of every teacher. Research has shown that metacognition is a contributing factor to student success and as metacognition is a core component in the Singapore mathematics curriculum framework as well, it is not surprising that local research has continued to pursue this strand of studies. Research involved framing metacognitive behaviours as well as the impact of metacognitive instructions for success. One of the studies on metacognitive instruction was carried out in a computer environment where students worked collaboratively. Research involving student sense-making in problem solving was also identified as another strand. Research studies include students' consideration of realities of the context of word problems, whether sense-making is gender related, and the effects of an intervention programme on the sense-making ability of Primary 3 students.

Research in the affective domain related to problem solving was also located. These research studies comprise determining the interrelationship between mathematics- anxious students and their performance in solving non-routine problems and students' perception in solving mathematics performance tasks. Finally, a strand on the use of ICT in problem solving included two studies. One is an intervention study where students used LOGO to solve open-ended problems and the other involved students using Knowledge Forum to negotiate meanings to discuss mathematical ideas for planning a budget.

The review provides a perspective on what the students are capable of, their difficulties, and what can be done to enhance their problem-solving endeavours and mathematics learning better. In a sense, while the outcomes of the research are valued and called for positive change, it must be noted that problem solving still depends on the role teachers take in the problem-solving paradigm as they engage the students. It is noted that the research carried out are aligned to curriculum goals since Singapore has a centralized education system. While the studies have contributed to adding to the knowledge of problem-solving research involving students in Singapore, there is always more that can be done to deepen the understanding in each strand through following-up or extending current research, and to find ways to maximise the outcomes of the research with stakeholders. As it stands, more research can be carried out in understanding students' construction of knowledge (their inventive resources), design of productive learning environments, building teacher capacity as well as in addressing recommendations and concerns raised in the research carried out. As the history of research in mathematics education in Singapore is relatively young, it must be acknowledged that the repertoire of studies cannot cover the field of mathematical problem solving involving students adequately. Nevertheless, it translates to more opportunities for local research to cover more ground in mathematical problem solving in the light of educational reform efforts.

CHAPTER 41

RESEARCH ON SINGAPORE MATHEMATICS CURRICULUM AND TEXTBOOKS

Searching for Reasons Behind Students' Outstanding Performance

Yan Zhu
East China Normal University, China

Lianghuo Fan
University of Southampton, UK

EXTENDED ABSTRACT

Over the last 15 or so years, Singapore mathematics curriculum and textbooks have attracted a considerable amount of attention internationally, particularly from educational researchers, reformers and policy makers. This phenomenon is closely related to people's interest in searching for the reasons to account for Singapore students' outstanding performances in well-referred international comparative studies, such as the Trends in International Mathematics and Science Study (TIMSS) and the Programme for International Student Assessment (PISA). This chapter is intended to provide readers with an overall look at the development of Singapore mathematics curriculum and textbooks and researches that have been carried out over the last decades on these two related areas, and discuss relevant issues concerning mathematics curriculum and textbook development.

The first section of the chapter starts with the historical background of the Singapore mathematics curriculum, which is followed in the second section with a brief description about the modern general education system in Singapore.

It has been said that the modern history of Singapore began in 1819 when it came under British colonial rule until 1959. The then education was mainly left to the different communities, which resulted in the co-existence of different schools, including Chinese schools, English schools, Malay schools, and Tamil schools. These schools not only used different languages as the medium of instruction, but also adopted different syllabi and textbooks, mostly imported from the US and the UK.

In 1959, Singapore attained self-government, and the first localized school syllabus in mathematics was implemented in 1959. It adopted a spiral approach and treated mathematics as a unified subject rather than a "many-branched" discipline. The second part of this syllabus was commonly known as Syllabus B. It was followed by Syllabus C in late 1960s, which witnessed the influence by the progressive new math movement starting in the US. Syllabus D started being implemented from the late 1970s, which to some extent reflected the back-to-basics trend. By the early 1990s, Singapore had reached a kind of equilibrium state for syllabi and textbooks with both being designed locally. In particular, problem solving has since then become the central theme in both primary and secondary mathematics curriculum, and problem solving heuristics and metacognition received highly explicit attention in the curriculum. Figure 42.1 shows the latest version of the well-known pentagonal curriculum framework.

The third section of the chapter provides a review of research literature on Singapore's mathematics curriculum with some notably being from comparative perspectives. These researches were closely related to Singapore students' outstanding performance as consistently demonstrated in international comparison studies. In general, researchers found that Singapore curriculum, which has historical connection with both West and East traditions, has been carefully planned, coherently developed, and effectively implemented. It placed clear emphasis on problem solving by making it the centre of the curriculum framework, created more differentiated learning for different students by streaming, and provided effective guidance and support for teachers and students to achieve the curriculum goals.

The fourth section focuses on the research about Singapore's mathematics textbooks, which were also often believed to be one important reason for Singapore students' excellent performance in the international comparisons. The studies conducted in this area have mainly included textbook analysis, textbook comparison, and textbook use.

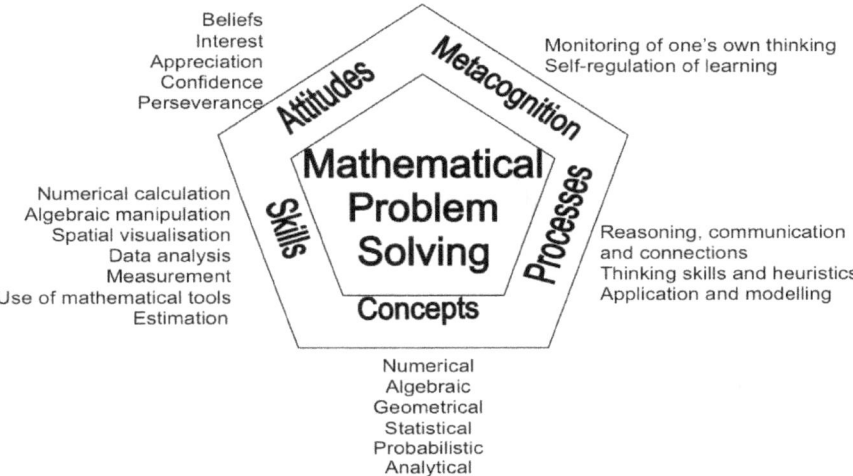

Figure 41.1. Singapore mathematics curriculum framework (Ministry of Education, 2006a, 2006b).

Overall, the researchers found that Singapore mathematics textbooks emphasized fundamental knowledge and skills, provided clear instruction on specific, though not all that the syllabus required, problem solving heuristics and procedures, and presented more challenging mathematics problems. Nevertheless, they are more traditional in orientation, or in other words, can hardly be called reformed textbooks. Moreover, in connection with the fact that schools in many countries have imported Singapore's mathematics textbooks for classroom use (e.g., see Quek, 2002), researchers have found that a successful adoption of another nation's textbooks took much more than simply using the textbooks (e.g., Ginsburg, Leinwand, Anstrom, & Pollock, 2005).

It is also noted from the available research literature that there were gaps between Singapore's national mathematics curriculum and its mathematics textbooks, which was particularly evident is textbooks' representing problem types and problem solving procedures (e.g., Fan & Zhu, 2000; Ng, 2002). The chapter argues that it will be interesting as well as important to further investigate why there were such gaps and how the curriculum policy makers and textbook developers can work together so textbooks can be better aligned with the curriculum.

The chapter ends with the following observation, that is, although many researchers conducted research on Singapore mathematics curriculum and textbooks because it is supposedly an important factor for its students' excellent achievement as revealed in international comparisons, there has been virtually no specific research that directly aims to address if there is direct or causal relationship between the mathematics curriculum/textbooks and students' achievement in Singapore, or more generally in any other countries. Therefore, this category of curriculum and textbook research remains much needed in order to obtain more scientifically reliable conclusion for curriculum reform and textbook development, though the method to conduct such research is likely more challenging (also see Fan, 2011).

REFERENCES

Fan, L. (2011, October). *Textbook research as scientific research: Towards a common ground for research on mathematics textbooks*. Paper presented at the International Conference on School Mathematics Textbooks. Shanghai, China.

Fan, L., & Zhu, Y. (2000). Problem solving in Singaporean secondary mathematics textbooks. *The Mathematics Educator, 5*(1-2), 117-141.

Ginsburg, A., Leinwand, S., Anstrom, T., & Pollock, E. (2005). *What the United States can learn from Singapore's world-class mathematics system (and what Singapore can learn from the United States): An exploratory study*. Washington, DC: American Institutes for Research.

Ministry of Education. (2006a). *Mathematics syllabus* (Primary). Singapore: Author.

Ministry of Education. (2006b). *Mathematics syllabus* (Secondary). Singapore: Author.

Ng, L. E. (2002). *Representation of problem solving in Singaporean primary mathematics textbooks with respect to types, Pólya's model and heuristics*. Unpublished master thesis, Nanyang Technological, Singapore.

Quek, T. (2002). US, Malaysia, Thailand, Viet Nam, India, Pakistan, Bangladesh, Finland ... Now, Israel uses S'pore maths textbooks too. *The Straits Times*, Sept. 23, 2002, Singapore.

CHAPTER 42

TEACHERS' ASSESSMENT LITERACY AND STUDENT LEARNING IN SINGAPORE MATHEMATICS CLASSROOMS

Kim Hong Koh
Nanyang Technological University, Singapore

ABSTRACT

To be productive in the workplace and to be informed citizens in the 21st century, our knowledge-based society requires individuals to attain mathematical literacy, which denotes the ability to grasp the implications of many mathematical concepts, to reason and communicate mathematically, and to solve non-routine, real-world problems effectively using a variety of mathematical methods. This set of competencies has been given more emphasis in 21st century curricula and teacher education programs around the globe, including in Singapore. This chapter begins with a discourse on the importance of realigning assessment and curriculum to the desired learning outcomes of the 21st century in the teaching and learning of mathematics. Since the mathematics reform movement in the late 1990s, many mathematics educators have advocated a shift of focus from the drill-and-practice of basic mathematical concepts and procedural skills to students' active learning and understanding of complex mathematical concepts through non-routine problem solving, mathematical thinking and reasoning, communication, and making connections to the real world. This focus on learning mathematics with understanding, or mathematical literacy, has also been endorsed by the standards of the National Council of Teachers of Mathematics in the US (NCTM, 1989) and the earlier Cockcroft Report *Mathematics Counts* in the UK (Cockcroft, 1982). The Singapore mathematics curriculum has a similar focus on mathematical literacy as the desired outcome of learning mathematics and mathematical problem solving is placed at the core of the curriculum framework that serves as a guide for the improvement of mathematics instruction and assessment in Singapore schools. Although the Singapore mathematics curriculum framework is under the influence of the global mathematics reform, it is also reviewed and revised regularly to be well aligned with the new policy initiatives launched by the Singapore Ministry of Education (MOE). Changes in curriculum should go in tandem with the changes in assessment. As such, part of this chapter also includes a discussion on some of the key policy initiatives by MOE and their implications on Singapore mathematics curriculum and assessment.

Similar to the teaching professionals in the US and the UK, mathematics teachers in Singapore have been urged to shift their assessment practices toward the use of alternative forms of assessment (i.e., open-ended questions, performance tasks, projects, journals, and portfolios) that are more well aligned with the higher-order curricular goals of 21st century. Because the goal of mathematics is to enable students to learn mathematics with understanding, a good way to assess students' attainment of this important goal is through their flexible performances of understanding and authentic assessment or performance assessment is touted as a viable assessment method that can capture students' performances of understanding. A substantial portion of this chapter is devoted to a review and discussion of the empirical studies conducted in relation to teachers' assessment practices in Singapore's 21st century mathematics classrooms in general and the global mathematics reform movement in particular. Although there have been various initiatives over the past two decades to introduce performance assessments into mathematics teaching and learning in Singapore classrooms, most of these initiatives are on a small-scale basis. At the systemic level, there are concerns with respect to the misalignment between the intended mathematics curriculum and the enacted assessment. This is partly due to the teachers' lack of assessment literacy in designing and implementing high-quality performance assessments. Therefore, effective teacher professional development in assessment literacy at the systemic level is of paramount importance.

Two key empirical studies on teacher professional development in designing and implementing performance assessments in mathematics were used in the latter part of the chapter to illustrate the importance of providing teachers with effective professional development. Such efforts are deemed important to improve teachers' assessment practices, which in turn will lead to higher quality of student learning in mathematics classrooms. The findings of the two studies indicate that there is a need for more systematic in-service training or professional development for mathematics teachers to improve their assessment literacy. Based on the seven features of effective professional development by Garet, Porter, Desimone, Birman, and Yoon (2001), the chapter concludes with a list of recommendations for the preparation of mathematics teachers in teacher education and professional development programs. The seven features of effective professional development include type of activity, duration, collective participation, focusing on content, active learning, fostering coherence and communication, and teacher outcomes.

CHAPTER 43

A THEORETICAL FRAMEWORK FOR UNDERSTANDING THE DIFFERENT ATTENTION RESOURCE DEMANDS OF LETTER-SYMBOLIC VERSUS MODEL METHOD

Swee Fong Ng
*National Institute of Education,
Nanyang Technological University, Singapore*

ABSTRACT

Representation is at the heart of mathematics. Because different representations illustrate different aspects of a particular idea it is therefore important for students to learn what representations to use and when to use them (NCTM, 2000, p. 69). More importantly students need to learn to interrogate what ideas are preserved by different representations and what new connections are presented by their alternatives (Kaput, 1987). Abstraction plays an important part in the construction of representations. Once abstracted symbols are then used to construct the representation. Because symbols make it easier to manipulate these ideas captured in the representation the decision as to what symbols to use is then crucial. The choice of symbols reflects the sensitivity to the cognitive readiness of learners to engage with these symbols. As students' representational stockpile increases, so must their ability to discern the strengths and limitations of each representation. When confronted with a problem their capacity to discern which representation is best suited for a specific problem reflects how the student understands the problem. Their choice of representation may shed light on new understandings acquired by students.

The Singapore mathematics curriculum puts a premium on students' capacity to solve word problems (Ministry of Education, 2001, 2006). To cultivate this capacity, students are taught to use various problem-solving heuristics. The policy to introduce problem-solving heuristics at the primary level has shown positive results. The heuristic 'draw a diagram' known locally as the model method seemed to be particularly effective. With this heuristic, rectangles are the symbols used to represent numbers. The resulting schematic representation, known as the model, captures the information presented in the text of a problem. The fact that rectangles can be used to represent specific numbers or unknowns makes the model method a powerful representational tool. This is because the construction of the appropriate model drawing builds on pupils' understanding of the part-part-whole nature of number. A number can be partitioned into two or more smaller parts (Van de Walle, 2001). Hence a whole rectangle can be partitioned into two or more smaller rectangles, with the length of each smaller rectangle reflecting the size of a particular number. The model method can be used to solve arithmetic as well as algebraic type word problems which would normally require the construction and solution of a system of linear equations involving one or two variables.

What makes the model such a powerful tool? One possible explanation is that it provides an avenue to first describe the problem before the process of calculating for the solution begins. This process of "first-describing-and-then-calculating" (Post, Lesh & Behr, 1988) is one of the key features that make algebra distinct from arithmetic. Ng and Lee (2009) showed that Primary Five children without any prior knowledge of letter-symbolic algebra but who were competent with the model method could solve algebraic word problems that would normally require the construction of linear algebraic equations up to two unknowns.

Their progress through the curriculum meant that these pupils as novice students of letter-symbolic algebra were expected to construct a system of linear equations to represent the information presented in algebra word problems. They were then expected to apply the relevant procedural skills and conceptual knowledge to solve this system of linear equations, skills and knowledge very distinct from those needed to solve for the unknown represented by the rectangles in the model method. There is rich body of research investigating issues related to using letters to construct algebraic objects such as expressions and equations to represent mathematical ideas and information (e.g. Kieran, 1989, 1981, 1997; Küchemann, 1981). In the Singapore curriculum with the introduction of each new representation - model method to algebraic equations - the symbols used vary in their complexity. Why is it that many students who were taught formal algebra continue to choose to use the model method to solve algebra word problems? What could students' choice of methods tell us about their understandings of the representation and the related choice of symbols and hence the method used to solve algebra word problems?

To answer the above questions, this chapter reviewed a suite of studies conducted in Singapore which investigated how primary pupils used the model method to solve algebraic word problems. I then investigated the strategies adopted by 124 Secondary 2 (14+) students from five schools to solve ten algebra word problems. The findings showed that although the Singapore model method was introduced as a heuristic to support the problem solving capacity of upper primary pupils, it has been found to impact on their subsequent acquisition of letter-symbolic algebra. This is consistent with recent neuroimaging studies (Lee, et al. 2007; Lee, et al. 2010) which demonstrated that letter-symbolic algebra drew more heavily on the attention resources of the students but which were unable to account for the difference between the two methods. This chapter provides a theoretical framework which explains all of these results, providing teachers with a deeper understanding of the difference between the two methods, in order to guide students through a more confident transition from model to algebraic solution methods.

REFERENCES

Kaput, J. J. (1987). Representation systems and mathematics. In C. Janvier (Ed.) *Problems of representation in the teaching and learning of mathematics* (pp. 19 – 26). Hillsdale, NJ: Lawrence Erlbaum Associates, Inc.

Kieran, C. (1989). The early learning of algebra: A structural perspective. In S. Wagner & C. Kieran *Research issues in the learning and teaching of algebra* (pp. 33 – 53). Reston, Virginia: NCTM.

Kieran, C. (1981). Concepts associated with the equality symbol. *Educational Studies in Mathematics, 12,* 317 -326.

Kieran, C. (1997). Mathematical concepts at the secondary school level: The learning of algebra and functions. In T. Nunes & P. Bryant (Eds.) *Learning and teaching mathematics: An international perspective* (pp. 133 - 158). East Sussex, UK: Psychology Press.

Küchemann, D. (1981). Algebra. In K. Hart (Ed.) *Children's understanding of mathematics: 11 – 16*. London: John Murray.

Lee, K., Lim; Z. Y., Yeong, S. H. M., Ng, S. F.; Venkatraman, V., Chee M. W. L. (2007). Strategic differences in algebraic problem solving: Strategic differences in algebraic problem solving: Neuroanatomical correlates. *Brain Research*, 1155 (June), 163 – 171.

Lee, K., Yeong, S. H. M., Ng, S. F., Venkatraman, V., Graham S., Chee, M. W. L. (2010) Computing Solutions to Algebraic Problems Using a Symbolic Versus a Schematic Strategy. *ZDM - The International Journal on Mathematics Education*, 42, 591 – 605.

Ministry of Education (2001). *Mathematics Syllabus Primary*. Singapore.

Ministry of Education (2006). *Mathematics Syllabus Primary*. Singapore.

National Council of Teachers of Mathematics (2000). *Principle and Standards for School Mathematics*. Reston, VA: Author.

Post, T. R.; Behr, M. J.; Lesh, R. (1988): Proportionality and the development of prealgebra understandings. In: A. Coxford; A. Shulte (Eds.), *The ideas of algebra, K-12*. Reston, VA: NCTM (1988 Yearbook), p. 78-90.

Van de Walle, J. A. (2001). *Elementary and middle school mathematics: Teaching developmentally - 4th edition*. New York: Addison Wesley Longman.

CHAPTER 44

A MULTIDIMENSIONAL APPROACH TO UNDERSTANDING IN MATHEMATICS AMONG GRADE 8 STUDENTS IN SINGAPORE

Boey Kok Leong, Shaljan Areepattamannil, and Berinderjeet Kaur
*National Institute of Education,
Nanyang Technological University, Singapore*

EXTENDED ABSTRACT

The performance of Singapore's grade 4 and 8 students on the Trends in International Mathematics and Science Study (TIMSS) assessments in 1995, 1999, 2003 and 2007 was outstanding. Despite the stellar performance of Singapore's students on international mathematics and science assessments, there is sparse research on these students' approaches to mathematical learning. Although there are different approaches or perspectives to understand and measure the mathematical learning of students in elementary and secondary schools, one approach that appears particularly promising and pertinent for the study of students' mathematical learning is the SPUR approach (see Thompson, Kaur, & Bleiler, 2010; Thompson & Senk, 2008). To ensure that both elementary and secondary students experience the multiple aspects of understanding mathematical situations, the University of Chicago School Mathematics Project (UCSMP) propounded the SPUR approach to mathematical learning (Usiskin, 2003; Viktora et al., 2008). According to the SPUR approach to mathematical learning, there are four important dimensions to understanding in mathematics: skills—having ways to obtain and check solutions; properties—knowing properties that identify and justify reasoning; uses—recognizing situations in which the mathematics is applied; and representations—picturing or otherwise representing the mathematical ideas (Viktora et al., 2008).

Only a small body of research, however, has examined the relationships between the SPUR objectives and students' achievement in mathematics (e.g., Thompson et al., 2010; Thompson & Senk, 2001, 2008). Studies that explored the relationships between the SPUR objectives and students' achievement in mathematics suggested that nurturing students' multi-dimensional approaches to understanding in mathematics might help them to develop a robust understanding of mathematical concepts and reasoning (Thompson et al., 2010; Thompson & Senk, 2008). Given the dearth of research examining the mathematics achievement of elementary and secondary students in terms of the SPUR approach to mathematical learning, more empirical research is warranted to evaluate the effectiveness of the SPUR approach to mathematical learning. Further, there is sparse research on the impact of student-leval and school-level factors on the SPUR approach to mathematical learning. Hence, the purpose of the present study was two-fold: first, to examine whether the mathematics achievement of grade 8 students in Singapore, who took part in TIMSS 2007, varied across the multiple dimensions of understanding in mathematics—skills, properties, uses, and representations (SPUR); and second, to investigate the predictive effects of student-level and school-level factors on grade 8 students' multiple dimensions of understanding in mathematics. Specifically, the following two research questions addressed the purpose of the study:

1. To what extent does the mathematics achievement of grade 8 students in Singapore, who participated in TIMSS 2007, vary across the multiple dimensions of understanding in mathematics, namely skills, properties, uses, and representations?

2. How well do student-level and school-level factors predict grade 8 students' multi-dimensional mathematical knowledge—skills, properties, uses, and representations—in Singapore?

To facilitate comparison with Singapore, we selected countries such as Australia, Chinese Taipei, England, Hong Kong-SAR, Japan, Korea, Minnesota, Massachusetts, and the United States that formed a comparison group. A mean score from each dimension of understanding was computed from the comparison group and t-test was used to determine if there was significant difference between the means of Singapore and the group. Countries were selected on the basis of their performance on the TIMSS 2007 assessment. The countries selected for the comparison group in the present study were those that scored higher than the international average of 500 in mathematics. Besides, the comparison group included some of the top performing East Asian countries in TIMSS 2007 or English speaking countries. A total of 88 released mathematics items from TIMSS 2007 were used in this study in which each item was categorised into exactly one of the four dimensions of understanding. The items comprised both multiple choice items and questions requiring constructed response worth up to 2 points. The student achievement was computed using the partial credit item response theory (IRT) model for multiple responses. IRT scales were created for every domain of understanding on a scale with mean 500 and standard deviation 100.

To address the first research question, correlations and t-tests were conducted. Singapore students' mean score in each dimension was statistically significantly higher than the mean from the comparison group. The amount of homework has a positive effect on score from each dimension of understanding. Similarly, positive affect toward mathematics, valuing mathematics, self-confidence in mathematics, home language (language of test), and school climate were positively correlated with the multiple dimensions of understanding in mathematics.

To answer the second research question, multilevel modelling analyses were conducted. The student-level variables were gender, home language, positive affect toward mathematics, valuing mathematics, self-confidence in mathematics, and frequency of mathematics homework. The school-level variables were school climate and school resources. Gender had negative predictive effects on skills, properties, and representations. However, there were no gender differences with regard to uses. Home language, positive affect toward mathematics, and self-confidence in mathematics had positive predictive effects on all dimensions of understanding in mathematics. While valuing mathematics had positive predictive effects on representations, it had no predictive effects on skills, properties, and uses. Whereas school climate had positive predictive effects on all dimensions of understanding in mathematics, school resources had no predictive effects on any of the dimensions of understanding in mathematics.

The findings in this study led us to a deeper appreciation and understanding in aligning the Singapore mathematics curriculum along the line of the SPUR approach. Greater emphasis on the implementation of such approach by teachers in the classroom could be considered. Future works could be research into the pedagogy for teachers to incorporate SPUR into the teaching of mathematics.

REFERENCES

Thompson, D. R., Kaur, B., & Bleiler, S. (2010, August). Using a multi-dimensional approach to understanding to assess primary students' mathematical knowledge. In Shimizu, Y., Sekiguchi, Y., & Hino, K. (Eds.) *Proceedings of the 5th East Asia Regional Conference on Mathematical Education, 2*, 472-479.

Thompson, D. R., & Senk, S. L. (2008). *A Multi-Dimensional Approach to Understanding in Mathematics Textbooks Developed by UCSMP.* Paper presented in Discussion Group 17 of the International Congress on Mathematics Education. Monterrey, Mexico.

Usiskin, Z. (2003). A personal history of the UCSMP secondary school curriculum: 1960-1999. In Stanic, G. M. A., & Kilpatrick, J. (Eds.), A history of school mathematics, Volume 1 (pp. 673-736). Reston, VA: National Council of Teachers of Mathematics.

Viktora, S. S., Cheung, E., Highstone, V., Capuzzi, C. R., Heeres, D., Metcalf, N. A., Sabrio, S., Jakucyn, N., & Usiskin, Z. (2008). The University of Chicago School Mathematics Project: Transition Mathematics. Chicago, IL: Wright Group/McGraw Hill.

MALAYSIA

SECTION EDITORS

Chap Sam Lim
Universiti Sains Malaysia

and

Bharath Sriraman
The University of Montana

Editorial Board Members

Liew Kee Kor, *Universiti Teknologi MARA Kedah*
Cheng Meng Chew, *Universiti Sains Malaysia*

CHAPTER 45

MATHEMATICS EDUCATION RESEARCH IN MALAYSIA

An Overview

Chap Sam Lim, Parmjit Singh,
Liew Kee Kor, and Cheng Meng Chew

ABSTRACT

This chapter aims to provide an overview and historical development of math education research in Malaysia since 1970. Research studies in math education in the earlier years were generally limited to postgraduate thesis and dissertations. The focus was mainly on cognitive development and teaching approaches in the 1970s. The research trend become more diverse in the 1980s with various topics ranging from attitudes, problem solving to error analysis. In the 1990s, the research areas continue to expand those topics found in the 1980s but with a significant focus on scheme of mathematical concepts. However, there were also some studies related to assessment and evaluation. Starting the year 2000, several new research areas such as language and mathematics; beliefs in mathematics; cultural differences and technology were explored. With the setting up of research universities, more and more research grants were awarded to the local researchers and educators. Consequently many more research projects were undertaken with the majority of the topics focusing on mathematical software and courseware. Generally there was a paradigm shift in research design from scientific to interpretative. Likewise, there was a change in choice of respondents from students to teachers or pre-service mathematics teachers. In order to provide a more comprehensive scenario of the research trend mentioned above, the chapter will end with an outline and brief introduction of the various chapters included in the Malaysian section.

Keywords: history of mathematics education, Malaysia, Malaysian trends in mathematics education research

CHAPTER 46

RESEARCH STUDIES IN THE LEARNING AND UNDERSTANDING OF MATHEMATICS

A Malaysian Context

Parmjit Singh and Sian Hoon Teoh

ABSTRACT

This chapter provides a critical and comprehensive review of mathematics learning and understanding studies that have been carried out in Malaysia. Four main themes were identified and discussed, namely schemes of learning in mathematical concepts; understanding and conception of mathematical terms and knowledge; development of concepts in mathematics instruction; and performance and achievement in mathematics. This will be followed by a detailed discussion of a study on multiplicative thinking in proportion and ratio. The implications and suggestions for future research will conclude the chapter.

Keywords: mathematics education studies, mathematical concept development, learning schemes

CHAPTER 47

NUMERACY STUDIES IN MALAYSIA

Munirah Ghazali and Abdul Razak Othman

ABSTRACT

The chapter comprises three sections: (a) issues on definition of numeracy; (b) numeracy in Malaysian curriculum and (c) numeracy studies in Malaysia. One of the critical issues is the definition of numeracy. This first section discusses different views on the existing definitions of numeracy and its related components. The second section reviews Primary School Mathematics Curriculum Specifications (in the Year 2002-2007) with regard to the aspects of numeracy in all topics. The third section reports and highlights all studies on numeracy and numeracy related areas in Malaysia.

Keywords: numeracy, numeracy in Malaysia

CHAPTER 48

MALAYSIAN RESEARCH IN GEOMETRY

Cheng Meng Chew

ABSTRACT

This chapter has four main objectives. The first objective is to discuss the importance of geometry in general and its implications for the Malaysian primary and secondary mathematics curricula in particular. The second objective is to provide a critical and comprehensive literature review of mathematics education studies in geometry that have been carried out in the country. The third objective is to present a detailed discussion of a particular study conducted by the author on the teaching and learning of geometry in a Malaysian classroom. The last objective is to offer suggestions for future research in geometry in the country.

Keywords: geometry, Malaysia primary and secondary mathematics curricula, teaching and learning of geometry

CHAPTER 49

RESEARCH IN MATHEMATICAL THINKING IN MALAYSIA

Some Issues and Suggestions

Shafia Abdul Rahman

ABSTRACT

The objective of this chapter is to provide an overview of research studies related to mathematical thinking that have been carried out in Malaysia. It begins with a brief introduction of the importance of mathematical thinking in the Malaysian school curriculum, followed by a critical review of the relevant local literature to shed some light on the extent to which these research works have contributed to the development of mathematical thinking in Malaysia. A critical discussion is presented on what constitutes mathematical thinking, drawing upon a recent research done by the author. The chapter concludes with some implications and suggestions for future research in mathematical thinking.

Keywords: mathematical thinking, Malaysian studies in mathematical thinking

CHAPTER 50

STUDIES ABOUT VALUES IN MATHEMATICS TEACHING AND LEARNING IN MALAYSIA

Sharifah Norul Akmar Syed Zamri and Mohd Uzi Dollah

ABSTRACT

This chapter describes some of the studies about values in mathematics teaching and learning in Malaysia. Studies of values in mathematics education in Malaysia can be considered quite sporadic and still in the expansion stage. The discussions in this chapter include development of values education in Malaysian schools, values in the Malaysian mathematics curriculum, problems and challenges in inculcating values in mathematics education and some studies that has been conducted regarding values in Malaysia. Studies conducted by researchers from three different universities were also discussed. The chapter wraps up with a detailed discussion of a study to explore values in mathematics teaching as espoused by three secondary schools mathematics teachers in Malaysia.

Keywords: beliefs versus values, values, values in mathematics teaching and learning

CHAPTER 51

TRANSFORMATION OF SCHOOL MATHEMATICS ASSESSMENT

Tee Yong Hwa, Chap Sam Lim, and Ngee Kiong Lau

ABSTRACT

This chapter begins with a brief historical development of Malaysian examination system. This is followed by the school mathematics curriculum and assessment in Malaysia and issues related to mathematics assessment. Next, a review of local literatures done on mathematics assessment will be presented. To elaborate further, a study that focuses on implementing a Mathematical Thinking Assessment (MaTA) Framework to assess students' thinking processes and its limitations will be discussed. This chapter concludes with some suggestions for future research of the study that will help to assess students learning in a more holistic and reliable fashion.

Keywords: examination system, Malaysia, assessment in Malaysia, mathematics curriculum in Malaysia, MaTA

CHAPTER 52

MATHEMATICS INCORPORATING GRAPHICS CALCULATOR TECHNOLOGY IN MALAYSIA

Liew Kee Kor

ABSTRACT

Graphics calculator technology encompasses more than just plotting of graphs. It features a variety of mathematical computational skills from manipulating symbolic expressions to manoeuvring analytic mathematical problems and equip with programming capabilities. This chapter begins with an overview on the use of graphics calculator in the teaching and learning of mathematics in Malaysia. In particular, it reports chronologically the stages of implementation of graphics calculator in the Malaysian mathematics curriculum. Consequently, it presents the findings of local research studies and gives an account on graphics calculator related events in Malaysia from year 2002 onwards.

Keywords: technology in mathematics instruction, graphing calculators, Malaysian mathematics curriculum

CHAPTER 53

MATHEMATICS TEACHER PROFESSIONAL DEVELOPMENT IN MALAYSIA

Chin Mon Chiew, Chap Sam Lim, and Ui Hock Cheah

ABSTRACT

This chapter aims to provide an overview of research on the professional development of mathematics teachers in Malaysia. Literature review on teacher professional development [TPD] indicates that the nature of TPD in Malaysia can be divided into two main strands: a) top-down teacher professional development programme initiated and conducted by the Ministry of Education through in-service courses or workshops with the objective to introduce innovations in teacher's teachings or prepare teachers for curriculum change; and b) research based TPD initiated by researchers with an intention of local interest to improve teaching practices such as Action Research and Lesson Study. Hence, the discussion in this chapter consists of two main parts. First, the situational context of mathematics teacher professional development will be elaborated to provide the background and setting. Second, the research findings on Action Research and Lesson Study in Malaysia will be discussed, primarily to examine their feasibility as an innovative form of teacher professional development. Finally, implications and future issues related to mathematics teacher professional development will also be addressed.

Keywords: teacher professional development, Malaysia, action research and lesson study

JAPAN

SECTION EDITOR
Yoshinori Shimizu
University of Tsukuba

Editorial Board

Yasuhiro Sekiguchi, *Yamaguchi University*
Keiko Hino, *Utsunomiya University*

CHAPTER 54

MATHEMATICS EDUCATION RESEARCH IN JAPAN

An Introduction

Yoshinori Shimizu
University of Tsukuba

SUMMATIVE ABSTRACT

This chapter intends to provides a birds-eye view on the areas of mathematics education research in Japan discussed in this section. The author emphasizes that we need to take into account the following contexts to understand the trends and issues in mathematics education research in Japan. First, in Japan we have national curriculum standards (the Course of Studies), which have been revised roughly every 10 years. In order to examine the trends and issues in most areas of mathematics research in Japan, we cannot neglect their connections with the goals and emphases described in the national curriculum standards. Second, the mathematics education community in Japan has a long tradition of lesson study by school teachers as practical research methodologies in the form of action research. Researchers and classroom teachers work closely within the community with local theories of students' learning in their perspectives. Then, to grasp the ongoing research agendas, we also need to pay careful attentions to their accumulated findings with respect to teaching materials and ways of teaching and learning in each research area. Third, developments of mathematics education research in Japan have been influenced by Western educational theories in various areas of inquiry, while educational activities themselves are rooted in East Asian cultural tradition. Thus, we need to look into the influences from both the East and the West in this regard.

The chapter then invites a historical perspective on mathematics education research in Japan (Chapter 56) and a discussion of developments of mathematics education research as a scientific inquiry (Chapter 57). These two chapters are mutually related and they together provide the warp and weft of the research in mathematics education in Japan. Then, the author identifies several key research areas that have been studied actively in recent years with historical backgrounds of efforts by teachers to tackle with long-standing issues in Japanese mathematics education; proportional reasoning (Chapter 58), students' understanding of algebra and use of literal symbols (Chapter 59), proof and proving as an explorative activity (Chapter 60), teaching and learning problem solving (Chapter 61), use of ICT tools in mathematics classrooms (Chapter 62), the role of metacognition in learning mathematics (Chapter 63). Then, two chapters follow with a focus on teaching and learning as cultural activities; cross-cultural studies on classroom practices (Chapter 64), and professional development of teachers through lesson study (Chapter 65).

The author points out again that for better understanding of accumulated research findings in mathematics education in Japan we need to see how those issues are identified in the community of teachers and how research again are conducted based on the efforts by teachers. For example, research on proportional reasoning in the Japanese context has a historical context of efforts through lesson study in the community of teachers to explore how and in what extent a teacher can provide a learning opportunities for the students to reason proportionally by using particular teaching materials and setting particular tasks. Thus, lesson study can be considered not only as the way professional developments are conducted, as discussed in Chapter 65 but also as a particular approach to research on teaching and learning.

CHAPTER 55

A HISTORICAL PERSPECTIVE ON MATHEMATICS EDUCATION RESEARCH IN JAPAN

Naomichi Makinae
University of Tsukuba

EXTENDED ABSTRACT

This chapter illustrates the history of mathematics education research as a field of academic inquiry in Japan. Makinae argues that it is generally difficult to point out exactly when mathematics education research has established and that we need to examine various phases of the process in which mathematics education research became a research field such as today through its historical development.

Japanese modern school system started in the early Meiji era (1868-1912) and a new school system named "Gaku-sei [School System]" was started by the government in 1872. In "Gaku-sei" modern mathematics education started in Japan. Thus, historically mathematics education research started from building totally new mathematics education within new modern school system. Referencing to western mathematics education, we had to develop curriculum, textbooks, teaching methods, and teacher training system. For the developments, the selection of teaching contents was most important issue of mathematics education research at that time. Mathematicians in universities, professors in teacher training colleges and teachers in attached schools have worked for them.

From the end of Meiji era to World War II (1900's-1945), under the influence of the movement of improvement of mathematics education in early twentieth century in Europe and the United States, curriculum development and improvement of teaching method improvement were studied from practical view point. Overseas studies had been translated and introduced. At first as for the curriculum development, separation in mathematics was considered. Secondary, contents of geometry were concerned. Logical construction based on Euclid's Elements was not suitable for children's learning in geometry. Thirdly, for the improvement of teaching method in class room, the view point of child centered education introduced in this era. Teaching based on proper subjects in mathematics was not suitable for children's learning. Instead of such a teaching, "Seikatsu-Sanjutu [Life Arithmetic]" was proposed. In this new teaching method, teaching from child life experience was emphasized. In the progress of improvement of School Mathematics under the influence of the movement of Europe and the United States, we can see signs of academic research for mathematics education. The first teachers' society of mathematics education "Nihon Tyuto Kyoiku Sugaku-kai [The Mathematical Association of Japan for Secondary Education (MAJSE)]" was established in 1919.

In this era, we can find an argumentation of mathematics education with academic perspectives. It was "Sugaku Kyoiku Shi [History of Mathematics Education]"(Ogura, 1932). In this book, he considered history of mathematics education as a cultural consequence. He described mathematics education with the social backgrounds, educational system and the thoughts of famous philosophers and education reformers. In post war period when mathematics education gained one of the divisions in university, his book played a role of evidence that there had been an academic research in mathematics education.

After World War II, Japanese educational system had changed rapidly and dramatically. National teacher training colleges were reorganized into universities in postwar educational reform. Becoming one of the divisions in such universities, the division of mathematics education was required academic standing point in research field. Developing theories of teaching and learning of mathematic were major objects of research in the contexts. Professional researchers of mathematics education had positions in the universities. Moreover in 1953, graduate schools of education (master's course and doctoral course) were set up in Tokyo University of Education and Hirosima University. Then, the graduate schools of education (master's course) were set up in universities of education and faculties of education throughout Japan.

MAJSE renamed twice in 1943 and 1970. After first rename "Nihon Sugaku Kyoiku Kai [Japan Society of Mathematical Education (JSME)]", the society broadened the scope of study to entire range of school mathematics.

In 1961 new journal "Sugaku Kyoiku Gaku Ronkyu [Full Discussion of Mathematics Education Research]" started to publish. This journal was for articles on fundamental investigations and scientific researchers in mathematics education, aiming to develop academic research. Adding to annual meeting, new academic conference "Sugaku Kyoiku Ronbun Happyo Kai [Conference for Presentation of Academic Paper of Mathematics Education]" was hold in 1966. The conference was for researchers in universities. In 1970 JSME renamed "Nihon Sugaku Kyoiku Gakkai [Japan Society of Mathematical Education (JSME)]. In Japanese the changing "Kai" to "Gakkai" meant clarifying characteristics of academic society. JSME have been an academic society of mathematics education keeping connection with practical teaching.

In the process of establishing research society, theories of teaching and learning of mathematics were developed. At first, traditional historical research and theory of teaching material took place in research. The historical study was consisted of review of past mathematics education. Through such a view, the background and fundamentals of current education were considered and directed the policy of mathematics education. It was typical form of description that theories of teaching materials were written following historical study. Secondary from teaching perspective, various approaches to teaching mathematics were developed. For examples, the discovery method in 1970's, Axiomatic method by Sugiyama, and Open-ended approach by Shimada and Nohda in 1980's were developed. Thirdly, cognitive studies of students learning were conducted by adopting the framework and methodologies in cognitive science. The focus of these researches was on students' understanding and thinking. Fourthly in according with the development of ICT technologies, the using of them in mathematics education had become a new research field.

Keywords: history of mathematics education, Japan, Japanese educational system

REFERENCES

Ogura, K. (1932). *A history of mathematics education* [in Japanese]. Tokyo, Japan: Iwanami-Shoten.

CHAPTER 56

THE DEVELOPMENT OF MATHEMATICS EDUCATION AS A RESEARCH FIELD IN JAPAN

Yasuhiro Sekiguchi
Yamaguchi University

EXTENDED ABSTRACT

Mathematics education Research in Japan has, like in other countries, been influenced by many factors such as changes of society, educational policy, and developments of research in other countries and other sciences, and so on. The development of mathematics education as a research field in Japan therefore cannot be examined without taking these factors into considerations. Sekiguchi discusses how mathematics education has developed as a research field in Japan by focusing on the developments of (1) conceptions of research, (2) research methodologies, and (3) research problems.

There are at least three major sources of Japanese mathematics education research. First, despite of the strong influence from research of the United States, Japanese education research has a strong tradition which values European educational research since the beginning of the Modern Area. Second, the Lesson Study tradition values classroom action research. Mathematics education researchers have been often challenged by school teachers who are engaging in lesson studies whether research conducted by researchers are useful to improve mathematics teaching in schools. Third, Japanese education has been strongly guided by the national syllabus. Curriculum material studies flourishes all over the country every time the national syllabus reform was on the issue.

After World War II, influence of educational research of the United States became very distinct in many areas. Since in the United States the behaviorist psychology and statistical analysis played important roles in educational research, Japanese mathematics educators also came to use such research methodology. However, in Japan there appeared also many other kinds of study articles which belong to historical studies, philosophical studies, curriculum material studies, Piagetian clinical studies, classroom action research, and the like in mathematics education research. Thus, Japanese mathematics education researchers had looked for a variety of research methodology. In 1990s constructivist research and qualitative research methodology came to rapidly receive wide recognition in mathematics education in the world and they has come to be valued in mathematics education in Japan. Currently, in Japanese mathematics education research community, various issues have been discussed, and various methodologies have been employed. For any research methodology, how to enhance the validity and relevance of research is a very important issue. For survey research and experimental research, there have been the standard methods to establish their validity under the positivist paradigm. Since newly proposed methodologies under a new research paradigm have limited ways to convince their validity, we need to discuss the research methodology of mathematics education in order to develop various ways to enhance the quality of research.

From the tradition of lesson study, Japanese mathematics education research has been closely connected to the classroom practices. As a result, it has obtained relatively high relevance to classroom practice. Though this is a great advantage, this may have hindered methodological argument, and theoretical development of research in Japan. Japanese researchers of mathematics education need to reflect their own ways of doing research, and strike a right balance.

Keywords: history of mathematics education, Japan; mathematics education as research discipline, methodologies of mathematics education in Japan, Japanese lesson study

CHAPTER 57

RESEARCH ON PROPORTIONAL REASONING IN JAPANESE CONTEXT

Keiko Hino
Utsunomiya University

EXTENDED ABSTRACT

Research on proportional reasoning has been one of the key research areas worldwide for a long period of time. Proportional reasoning refers to reasoning, or making assertions, based on the proportional relationships between the two quantities that are varying together. Proportional reasoning relates to a variety of mathematical concepts, skills or ways of thinking. Some researchers view proportional reasoning as a watershed concept, as both the capstone of elementary school arithmetic and the cornerstone of higher-level areas of mathematics (Lesh, Post, & Behr, 1988). Consequently, it has close connections with different areas of elementary mathematics, such as fraction, long division, percent, measurement, ratio and rate, as well as algebra and function and other higher level areas of mathematics.

In Japan, teaching of ratio and proportion has also been a big interest in the community of mathematics education. In her chapter, Hino synthesizes a body of research on proportional reasoning in Japan from three perspectives: change of curricular emphasis and its relationship to research on proportional reasoning, connection between research and classroom practice on ratio and proportion, and current research agendas. With respect to the related mathematical concepts or skills, this chapter concentrates mainly on the areas of ratio and proportion at the elementary school level.

Ratio has been an important curricular content in Japanese elementary schools since Meiji era. However, goals of teaching ratio were different between pre- and post- World War II. It was changed from children's becoming well versed in calculation to developing ways of thinking that are important to understand functioning of numbers and meaning of multiplication and division and to grasp quantitative relationships. Hino describes the change of curricular emphasis after 1958 and the treatment of ratio and proportion in elementary school curriculum.

It was the 1958 Course of Study that the strand of *quantitative relationship* was developed together with the three strands of *number and calculation*, *quantity and measurement* and *geometrical figure*. The *quantitative relationship* strand consisted of three sub-strands, one of which was *wariai* in Japanese. *Wariai* is a colloquial way of expressing the multiplicative relationship of one quantity with the other quantity. It was decided to adopt *wariai* as the name of the strand because emphasis was placed on fostering children's ratio concept through the learning of different contents such as numbers and calculations, measurements, and geometrical figures. The *wariai* sub-strand included a variety of topics: meaning of number, meaning of multiplication/division, functional viewing and thinking, *buai* and percent, and ratio on the quantities of different kinds. In the 1968 Course of Study, owing to the mathematics modernization, mathematical thinking and the ideas of set, structure and function were emphasized. Many topics included in the *wariai* sub-strand were moved to different strands and as a result, the name *wariai* was changed to *function* by focusing on "functional viewing and thinking." In the 1977 Course of Study, in the effort of overall reduction of content toward basics, the teaching of ratio concentrated on the grades of five and six. In the 1989 Course of Study, together with the emphasis of cultivating children's attitude toward utilizing mathematics in daily life situations, advantages of mathematical ways of viewing and handling objects were stated. In the 1998 Course of Study, the term *mathematical activity* was included in the goal statement of mathematics education at the elementary school level. Mathematical activity continued to be the focal point of curriculum and instruction in the newest Course of Study that was released in 2008. These different emphases on curriculum have influenced the teaching of ratio and proportion in Japan.

Hino also summarizes accumulated findings on the teaching of ratio and proportion by practice-based researches. In Japan, we have a tradition of practice-based research by schoolteachers. This tradition contributes to the realization of close connection between research and classroom practice. One of the focuses that appear repeatedly is how to introduce the concept of ratio or proportion in the classroom. In the introduction phase of the teaching, open-ended problems (Becker & Shimada, 1997) have also been developed and implemented in the classroom. For example, in the introduction to proportion in the sixth grade, different situations in which two quantities are varying together are often provided to children. Children are asked to compare/contrast similarities and differences among the situations and to classify the situations according to increase or decrease of the quantities. Based on the classification, children are further led to study features of proportional relationship in the subsequent lessons. Ways of kneading children's different strategies toward the concept that is the target of the lesson are also major focus of research. For example, in the introduction to ratio on different quantities (population density), children's different ways of comparing crowdedness in the two rooms are anticipated by teacher. Teacher plans the lesson and studies how to knead children's different ways toward the concept of population density.

Next, Hino summarizes and identifies current trends in research into five categories. They are described as (1) Studying the curricular sequence on ratio and proportion; (2) Connection between proportion-related mathematical content in elementary and lower secondary education; (3) Development of teaching materials on multiplication/division and ratio; (4) Proportional reasoning from the perspective of mathematical literacy; and (5) Clinical approach to children's learning processes in the classroom. For example, in the category of "Connection between proportion-related mathematical content in elementary and lower secondary education," she describes several researches that pay attention to how to facilitate children's shift of focus from within-ratio to between-ratio, and researches that address the importance of activity of symbolizing proportional relationships. In the category of "Development of teaching materials on multiplication/division and ratio," researches that tries to make surface the proportional relationship that we took for granted and to teach children to make use of the relationship to solve the problem are described. In doing so, researchers put forward the use of number line to visualize proportional reasoning that children engaged in to find the answers for multiplication and division problems. These researchers argue that, although the number line has been recognized as an effective way of grasping quantitative relationships and giving reasons for the decision of appropriate arithmetical operations, it also provides children with opportunity to pay attention to mathematical structure of the situations. In the category of "Clinical approach to children's learning processes in the classroom," by catching the quality of children's thinking during the lesson, researches that try to improve classroom teaching of ratio and proportion from the perspective of the learner are shown. Researchers in this category usually choose several focused children and collect detailed data by using camera for each focus child throughout the lessons in a teaching unit. The results show children's proportional reasoning that is actually used and functioned to acquire the knowledge and skill in multiplication/division and ratio. For example, it shows those children's activities on unitizing and norming the quantities are various and that they are the keys for accounting for the progressions the children made. The results also point out the importance of interaction between children and their inscriptions. It is essential that children interact with the diagrams without losing their own interpretation of the problem situation, even though children's diagrams vary beyond the formal number line representation.

Based on the research trend described, in the last section, Hino makes some suggestions for future research in Japan. The first suggestion is the need of healthy merge of accumulated findings in a variety of research focus. Another tasks for the future is the investigation on the ability to reason proportionally in more diverse populations. Since ratio and proportion is mainly taught in 5-7 grades in Japan, the research tends to concentrate on children in these grade levels. The roles of teacher should also be more focused in the teaching of ratio and proportion in the classroom. It is the teacher who actually provides learning opportunities for children by using certain teaching materials. Therefore, we need to know more about the teachers' understandings and conceptions on proportional reasoning and how to assist them in realizing the classroom in which the children are actively engaging in and developing their proportional reasoning abilities.

Keywords: proportional reasoning, trends in research on ratio and proportion, Japan

REFERENCES

Becker, J. P., & Shimada, S. (Eds.). (1997). *The Open-Ended Approach: A new proposal for teaching mathematics.* Reston, VA: National Council of Teachers of Mathematics. (Original work published in 1977)

Lesh, R., Post, T., & Behr, M. (1988). Proportional reasoning. In J. Hiebert, & M. Behr. (Eds.), *Number concepts and operations in the middle grades: Research agenda for mathematics education* (p. 93-18). Reston, VA: National Council of Teachers of Mathematics.

CHAPTER 58

JAPANESE STUDENT'S UNDERSTANDING OF SCHOOL ALGEBRA

Toshiakira Fujii
Tokyo Gakugei University

EXTENDED ABSTRACT

Understanding of algebra in school mathematics is one of the most important goals for secondary mathematics education. On the other hand, algebra in curriculum may differ among countries. In fact, some may be surprised to see that there isn't an algebra strand in the Japanese Course of Study of elementary and even of lower secondary school mathematics. This fact is contrasting to the American curriculum recently released as the Common Core State Standards which includes "Operations and Algebraic Thinking" even from Kindergarten to Grade 5.

The Course of Study in Japan is an official document for curriculum that is published and revised in about every ten years by the Ministry of Education, Culture, Sports, Science, and Technology (hereafter referred to as the Ministry). Therefore if there is no strand concerning with algebra in the Course of Study, it is likely that there isn't a unit titled as "algebra" in the elementary and lower secondary mathematics textbook which need to be officially approved by the Ministry although the textbooks are commercially published.

Instead of being an independent domain in the curriculum, algebra is systematically included in various parts of the mathematics curriculum, particularly at elementary level. Watanabe has made a list by analyzing Japanese textbooks from a foreigner's eyes and identified examples to show much of what is considered algebra exists in Japanese elementary mathematics textbooks (Watanabe, 2008). The main source of "algebra" is the Quantitative Relations strand in the Course of Study. That strand consists of ideas of functions, writing and interpreting mathematical expressions, and statistical manipulations. The contents of the Quantitative Relations strand have strong relations with the other three strands, Numbers and Calculations, Quantities and Measurement, and Geometric Figures (Takahashi, Watanabe, and Yoshida, 2004, p.67). In other words, these strands also include ingredients of algebra and algebraic thinking. At lower secondary level, it is much easier to identify what is considered algebra.

Although there isn't an exact unit titled as "algebra" in the elementary and lower secondary mathematics textbook, understanding of algebra in school mathematics is an important goal for mathematics education. However, many reports identify specific difficulties of learning of algebra. From international perspective, difficulties are labeled such as the cognitive obstacles, lack of closure, name-process dilemma, letter as objects, misapplication of the concatenation notation, misinterpretation of order system in number and so on. Matz (1982) also has identified inappropriate but plausible use of literal symbols in the process of transforming algebraic expressions.

In Japan, we are facing with the same problem that many students in lower secondary school are still confusing unknown numbers and variables. However, we need to be careful of diagnosing of their nature of understanding, simply because students seem to be good at solving conventional school type problems. Although ratios of correct answers in mathematics achievement tests such as IEA results and PISA results are high, Japanese mathematics educators suspect that limited understanding may coexist with this apparent success story.

This chapter reports on accumulated research results in order to probe Japanese students' understanding of school algebra focusing on the literal symbols, based mainly on research papers published in the journals of Japan Society of Mathematical Education. Then, three detailed discussions follow: The first focuses on the conventions or rules in the context of expressing and interpreting of mathematical expressions. The second focuses on research results could probe the understanding lying behind Japanese students' apparent procedural efficiency. The third focuses on a recent and alternative way of teaching of school algebra. That is made to show how the curriculum of the elementary and also secondary school can offer better opportunities for young people to think algebraically. Utilizing the potentially algebraic nature of arithmetic is one way of building a stronger bridge between early arithmetical experiences and the concept of a variable (Fujii & Stephens, 2008). The terms generalisable numerical expressions or quasi-variable expressions are to use in this chapter to make a case for a needed reform to the curriculum of the elementary and secondary school.

Keywords: arithmetic versus algebra, learning of algebra, algebraic thinking, Japanese studies on algebraic thinking; quasi-variable

REFERENCES

Fujii, T., & Stephens, M. (2008). Using number sentences to introduce the idea of variable. In *Algebra and Algebraic Thinking in School Mathematics, Seventieth Yearbook*, Reston, VA: National Council of Teachers of Mathematics.

Matz, M. (1982). Toward a Computational Theory of Algebraic Competence. *Journal of Mathematical Behavior,* 3(1), 93-166.

Takahashi, A, T. Watanabe and M. Yoshida (2004). *Elementary School Teaching Guide for the Japanese Course of Study: Arithmetic (Grade 1-6).* Madison, NJ: Global Education Recourses.

Watanabe. (2008) Algebra in elementary school: A Japanese perspective. *Algebra and Algebraic Thinking in School Mathematics, Seventieth Yearbook*, Reston, VA: National Council of Teachers of Mathematics.

CHAPTER 59

PROVING AS AN EXPLORATIVE ACTIVITY IN MATHEMATICS EDUCATION

Mikio Miyazaki
Shinshu University

Taro Fujita
University of Plymouth

EXTENDED ABSTRACT

In this chapter, Miyazaki and Fujita discuss trends and issues in recent studies into the teaching and learning of proof in Japan and summarize their findings, outcomes and implications. First they describe the general background about the current teaching and learning of mathematics and proof in Japan. Then, they review and synthesize the findings of research studies which have been presented and published in the last 10 years in accordance with themes and methods, and give their critical review. Finally, based on their review, they identify originalities and features of Japanese research, and specify inferences and suggestions for future research from the viewpoint of "proving as an explorative activity". They consider these will be essential to improve the current situation of the teaching and learning of proof in Japan.

The authors start with an introductory summary of the current teaching of proof and proving in, the Japanese national curriculum, the so-called 'Course of study'. In particular, they describe when formal proof is introduced in junior high school, what mathematical content is used to teach proof and so on. The recent results from recent national surveys in Japan are shown to describe the current situation as regards to teachers' approach to teaching proof, students' understanding of proof, typical misconceptions and so on.

Then, in section 2 and 3, they classify reviewed research studies into 'global studies' and 'local studies.' They broadly examines global studies, by which they mean studies of an overarching nature, not restricted to any particular stages/activities etc. First, they take theoretical and philosophical issues related to proof and proving such as the nature of proof, historical issues related to the evolution of the curriculum concerning the learning of proofs, -issues arising from the current situation as regards students' understanding of proof (Fawcett, 1938; Shimizu, 1997; Sugiyama, 1986). Next they focus on studies based on empirical approaches related to the significance of proof (Kunimune, Fujita, & Jones, 2009), planning and construction of proof, discovery of new properties based on constructed proof, discrepancies between inductive explanations and deductive proofs.

They then examine more specific aspects of the teaching and learning of proof and proving, i.e. local studies. The related research studies are reviewed in terms of the following themes: the 'seeds' (the foundations and emergence of ideas of proof and proving such as operative proofs or action proofs); the developmental process (the transition processes between informal proof and formal proof as the next step.); students' proving activities (proving as investigative activities, producing propositions, producing proofs and looking back at proving processes and proofs); curriculum development (principles to develop students' logical thinking and so on).

Based on their critical review, in section 4, the authors consider that proving should be viewed as flexible, dynamic and productive in nature, and various aspects of proving activities are interrelated and resonant with each other. To conceptualize this idea, they propose their triangular model of "proving as an explorative activity" which has the following three key components: producing propositions, producing proofs (planning and construction) and looking back (interpreting, evaluation and improving). This model provides us with an idea of how to make the teaching of proof more valuable for teachers and students. The two main directions of future research studies can be considered within their conical model of the developmental aspects of "proving as an explorative activity". One direction has a 'horizontal' nature to undertake new research to consider how to achieve the teaching and learning of explorative proving within all areas of mathematics. Another has a 'vertical' nature to consider the value of such an approach surrounding the teaching of proof and proving in other school subjects and students' activity outside schools.

In conclusion, the pedagogical idea of "proving as an explorative activity" enables proving to become a rich mathematical activity for many students.

To achieve this, researchers coming with a theoretical perspective must establish a long-term relationship with teachers who have daily encounters with children and students, must identify and select truly important future tasks for mathematics education from issues identified in mathematics lessons, and must provide concrete and achievable plans to improve the status quo by dialectic approaches, i.e. critical interaction between theory and practice.

Keywords: proving, proof trajectories, teaching and learning of proof, Japanese studies in proof

REFERENCES

Fawcett, H.P. (1938). The *nature of proof: a description and evaluation of certain procedures used in a senior high school to develop an understanding of the nature of proof* (The National Council of Teachers of Mathematics, The Thirteenth Yearbook), AMS PRESS, New York.

Kunimune, S., Fujita, T. & Jones, K. (2009b) "Why do we have to prove this?" Fostering students' understanding of 'proof' in geometry in lower secondary school. Lin, F-L, Hsieh, F-J., Hanna, G. and de Villiers, M. (eds.), *Proof and proving in mathematics education ICMI study 19 conference proceeding* (Vol. 1, pp. 256-261). Taipei: National Taiwan Normal University.

Shimizu, S. (1997). Views on scholastic ability by Dr. D. Kikuchi and Dr. R. Fujisawa. In Japan Society of Mathematical Education (Ed.), *Mathematics Education in Japan 1996: Philosophies of Mathematics Education in the Twentieth Century* (pp.17-28), Tokyo: Sangyotosho.

Sugiyama, Y. (1986). *Teaching mathematics based on axiomatic methods*, Tokyo: Touyoukan. (ISBN-13: 978-4491025452) [In Japanese]

CHAPTER 60

DEVELOPMENTS IN RESEARCH ON MATHEMATICAL PROBLEM SOLVING IN JAPAN

Kazuhiko Nunokawa
Joetsu University of Education

EXTENDED ABSTRACT

Nunokawa overviews the researches on mathematical problem solving which have been carried out in Japan to demonstrate the outline of changes in those researches during these 5 decades. Problem solving approach is often used in Japanese classrooms to encourage our students to learn mathematics actively and understand mathematical knowledge deeply. Because such teaching practices in Japan and their features are already reported and analyzed well by other researchers, this chapter focuses on the researches rather than teaching practices implemented in Japanese classrooms. Problem solving and mathematical thinking has been emphasized in Japanese mathematics lessons since 1930s, and teachers' interventions which were closely related with the conceptions of solving processes were discussed in 1960s and 1970s. But when the researches on mathematical problem solving carried out in the US and Europe were introduced to the mathematics education community in Japan at the end of 1970s and the beginning of 1980s, those works, like Schoenfeld (1985) and Silver (1985), had a great impact on the researches on problem solving in Japan. Thus, this chapter mainly deals with the researches before this impact (from the mid of 1950s to 1970s) and after that (from 1980s to 2000s) to illustrate that the developments in the researches on problem solving after that impact can be considered a kind of integrations of that impact and the intrinsic features of the researches in Japanese mathematics education community.

The origin of these intrinsic or traditional features can be found in the textbook series, called "Green-Cover Textbooks," which was edited during 1930s. Naomichi Shiono, the official chief editor of that textbook series, adopted the disposition to think mathematically as one of the objectives of mathematics education in elementary schools. He expected children to develop this disposition to find mathematical aspects in various phenomena and think those phenomena mathematically as well as to inquire mathematical phenomena and appreciate their beauty. It is usually said that his emphasis on this disposition has been passed down and made an important aspect of mathematics education in Japan (e.g. Oku, 1982). For example, when the effect of diagrams similar to a semantic network in the schema theory was studied by some researchers during 1950s, these researchers did not confine their scope to simple one- or two-step word problems and thought that it was important for students to experience the processes in which they investigated the relationships in problems through completing these diagrams.

One of the most fruitful products in Japanese researches on problem solving before the above-mentioned impact is the "Open-End Approach" project which started in 1971. While his framework included activities similar to mathematical modeling and formulating problems, Shimada (1977) intended to assess students' performances concerning "higher-order objectives" or "the objectives of higher-order thinking" and observe how students drew on all of their mathematical knowledge to tackle with unfamiliar problems, i.e. students' mathematical thinking. In 1980s, "the idea to use some form of open-ended problems in the mathematics classroom spread all over the world, and research on its possibilities became very vivid in many countries" (Phekonen, 1997). Moreover, this project promoted developing mathematical problems which focused on higher order thinking and perspectives which included students' constructions of problems.

Partly because of the influence of this tradition, Japanese researchers did not accept the researches on mathematical problem solving conducted in the US and Europe uncritically (Oku, 1982) when those researchers were introduced to the Japanese mathematics education community at the end of 1970s and the beginning of 1980s. While some researchers highlighted the importance of mathematical natures of problem solving through the comparison of researches and teaching practices in the foreign countries and in Japan, some researchers tried to use the concepts used in the researches in the US and Europe to systematically investigate mathematical thinking which Japanese researchers had studied and emphasized without those concepts. Among these concepts were included schemata, heuristics or problem solving strategies, metacognition, beliefs and problem posing. Furthermore, many of the researches on problem solving in Japan which were carried out during 1990s and 2000s attempted to elaborate those concepts so that those concepts can have characteristics which reflect the mathematical natures of problem solving and the disposition to think mathematically more explicitly. In this sense, the researches which were conducted after the impact can be considered a kind of integrations of the impact of the US and European researches and the intrinsic features held in Japanese mathematics education community.

For example, there were researches on problem solving strategies which incorporated the mathematical values into their framework or which illustrated the role of strategies more consistent with the mathematical natures of problem solving. Some researches on metacognition focused on the metacognition which functioned at turning points in problem solving processes and played critical roles for improving students' problem solving mathematically. The conceptions of problem solving processes presented by those Japanese researchers stressed some kinds of changes in solvers' understandings of problems, and those conceptions were expected to comprehend problem-solving processes without loss of the natures of mathematical thinking.

We can easily assume that trying to keep mathematical-thinking taste in researches on problem solving enable us to apply the findings and the theoretical frameworks in those researches to other research areas which are closely related to students' thinking. In addition to this feature, as teaching and learning via problem solving is the standard approach to teaching all mathematics contents in Japan, it can be expected that some components of mathematical problem solving play important roles and these components are constructed and developed in mathematics classrooms. Shimizu (1995) investigated students' metacognition observed in learning of the division of fractions and discussed the relationship between their metacognition and their flexibility in learning. Hirabayashi & Shigematsu (1986) searched for the sources of students' metacognition in their classrooms and explored how students' metacognition is constructed through the teaching activities of their teachers. Some researchers utilized the results and the perspectives derived from problem solving researches in order to investigate students' learning processes in more detail focusing on students' making-sense activities. The researchers pointed out the importance of some kinds of changes in students' making-sense of the tasks and in students' drawings, and then they investigated how those changes occurred in students' learning and how those changes could promote their learning.

This chapter attempts to illustrate that the Japanese research on problem solving has been constantly affected by its tradition which values mathematical nature and mathematical richness of thinking and learning, even after Japanese researchers learned much about problem solving research from the US and European researches. As the above examples show, such an approach to the research on problem solving seems to make the followings easier: (a) to apply the viewpoints and the findings in the problem solving research field to the researches on students' learning and on lessons in which students are provided with opportunities of doing mathematics; (b) to transfer from the research on learning and lessons to the problem solving settings in order to investigate students' solving processes and mathematical thinking in wider contexts. This interactive relationship will be expected to make the researches in both fields more fruitful.

Keywords: problem-solving research, Japanese research in problem-solving, metacognition, problem-solving strategy, problem-solving process

REFERENCES

Hirabayashi, I. & Shigematsu, K. (1986). Meta-cognition: The role of "Inner Teacher." In L. Burton & C. Hoyles (Eds.), *Proceedings of the 10th Annual Meeting of the International Group for the Psychology of Mathematics Education* (pp. 165-170). London.

Oku, S. (1982). Mathematical problem solving and mathematics education in Japan (1). *Tsukuba Journal of Educational Study in Mathematics, 1*, 27-36. [In Japanese]

Pehkonen, E. (1997). The state-of-art in mathematical creativity. *ZDM, 29* (3), 63-67.

Schoenfeld, A. H. (1985). *Mathematical problem solving*. Orland, FL: Academic Press.

Shimizu, Y. (1995). Students' thinking on division of fractions: The rigidity in their argument. *Reports of Mathematical Education: Journal of Japan Society of Mathematical Education, 63/64*, 3-26.

Silver, E. A. (Ed.). (1985). *Teaching and learning mathematical problem solving: Multiple research perspectives*. Hillsdale, NJ: Lawrence Erlbaum Associates.

CHAPTER 61

RESEARCH ON TEACHING AND LEARNING MATHEMATICS WITH INFORMATION AND COMMUNICATION TECHNOLOGY

Yasuyuki Iijima
Aich University of Education

EXTENDED ABSTRACT

This chapter illustrates aspects of the research and practice about teaching and learning with ICT (information and communication technology; mainly computers and the Internet) in Japan. The first aspect is the software development, the second is the use of the Internet, and the third is the style of educational practice in the classroom. To illustrate these aspects in the use of ICT, Iijima focused on the particular software, Geometric Constructor (in short, GC), which is one of the popular dynamic geometry software in Japan.

Our mathematical investigation has been changed in the depth and width, using mathematical tool software. Mathematical tool software used in Japan has been consisted with four groups; function graph software (e.g. Grapes), dynamic geometry software (e.g. Geometric Constructor, Cabri,

Geometer's SketchPad), spreadsheets (e.g. Excel), and CAS (e.g. Mathematica, Risa/Asir).

Since 1990, the style of educational practice was changing according to the principles of Ministry of Education of Japan about ICT use in education, course of studies, developments of the software and hardware and the Internet in Japan.

After his review on research on teaching and learning with ICT technologies in classrooms, Iijima discusses implications and suggestions for future research as follows. (1) For effective use of educational software, it is important that we make the community of developer and users (teachers) and make a cycle of development – use – evaluation – revision with the discussion in the community. (2) For the use of educational software in many scenes, it is useful to develop the software as web application. To achieve it, we can use the technology of Java applet, Flash, html5 & JavaScript etc. for client side software and the technology of CGI, php or asp.net etc. for server side software. (3) In Japan, for the usual use of ICT in many classrooms, it is effective to make digital contents based on the usual textbook. They are developed as digital textbook for teacher's presentation, which is used with interactive whiteboard or projector and screen in the usual classroom. (4) Digital textbook for teacher's presentation show the standard of educational practice using ICT. And we can use the Internet to share various contents and ideas for lessons in our communities. (5) Iijima has developed GC/html5 and some contents with it since 2010, which can be used as web application with PC and tablets (iPad, Android), to realize the possibility of netbooks or tablets to make possible the individual and/or group investigations in usual classrooms.

Keywords: ICT, mathematics teaching and learning with ICT

CHAPTER 62

"INNER TEACHER"

The Role of Metacognition in Learning Mathematics and Its Implication to Improving Classroom Practice

Keiichi Shigematsu
Nara University of Education

EXTENDED ABSTRACT

In the 1970s, research on metacognition has been focused on reading and in the early 1980, research started to focus on meta-memory in mathematics education. At that time, researchers did not recognize metacognition as a key component of students' learning both in research and practice, and rarely discussed at Japanese national conferences. This paper will discuss the general overview on research about metacognition and the present research results on the teaching-learning processes in Japan.

About twenty years ago Hirabayashi and Shigematsu (1986) who recognized metacognition as an important driving force in learning mathematics proposed the concept of "inner teacher", by hypothesize the process of internalization of metacognition in seven steps. The term "inner teacher" was used for standing for the process of what the teacher tells to students becomes to play metacognitive functioning after internalizing. Their ultimate goal of our research is to develop a clear concept about the nature of metacognition and applying this knowledge to improve the methods of teaching mathematics and teacher education. Their unique concept is that this "metacognition" is thought to originate from and internalized by the teacher him/herself. Teachers can not teach any knowledge per se directly to students but teach inevitably through their interaction with students in class. Their original conception is that this executive ego is really a substitute or a copy of the teacher from whom the student learns. The teacher, if he/she is a good teacher, should ultimately transfer some essential parts of his/her role to the executive ego of the student. In this context, they refer to the executive ego or "metacognition" as "Inner Teacher".

In his chapter, Shigematsu overview the findings and methodologies of research on metacoginition in mathematics education research in Japan. Metacognition is to study the mechanism of students' mathematics and problem solving, and in the practice of teaching is an essential element. On the other hand, in the actual teaching tends to focus only likely to be easily visible aspects of cognition, thinking the relationship between metacognition and cognition is an important agenda for mathematics education and research. And, these metacognitive studies, research and practice is also to reform the student and teacher autonomy in mathematics education, so to speak, it is desirable future development of more and more research. Shigematsu examines methodologies used in research on metacognition in mathematics education; questionnaire, observations of collaborative problem solving, the use of stimulated-recall interview technique, and student's journal writing.

Shigematsu et al. (2002), in particula, proposed that the practical continuous use of everyday mathematics teaching "Math Journal Writing Method" by the proposed step is divided into four specific teaching steps aim to help metacognition in students for middle grades elementary schools. The possibility of making such a practice has been verified by a teacher who wrote mathematics teaching work for the first time. This method is putting teachers through continuous red pen has the characteristic effecting teacher's interaction with students. In other words, writing a single journal is or can be viewed through that continuously analysis that the changes in students can be recognized through the teachers comment.

Based on research described above, Shigematsu identifies two issues for further research on metacognition in mathematics education. First, we need to understand the development of metacognitive learning during students' problem solving in a classroom instruction. For example, the question of how students' metacognitive action can be facilitated during 'Neriage', which means a whole class discussion period. In addition to make explicit metacognitive training course content based on the previous research, it is desirable to demonstrate that the curriculum guidelines for mathematics education as one of the purposes of fostering metacognition In this regards, examining Singapore's mathematics curriculum framework would be helpful. Second, while metacognition is an essential element to study the mechanism of students' mathematics and problem solving, teaching practices tends to focus more on the visible aspects of cognition. So we must think that the relationship between metacognition and cognition is an important agenda for mathematics education and research. These metacognitive studies for practice may also improve reform the student autonomous learning in mathematics education. In the future, the more and more research and practice for metacognition will be desired from now on.

Keywords: metacognition, cognition, teacher education, Japanese research on metacognition

REFERENCES

Hirabayashi, I. & Shigematsu, K. (1986). Meta-cognition: The role of "Inner Teacher." In L. Burton & C. Hoyles (Eds.), *Proceedings of the 10th Annual Meeting of the International Group for the Psychology of Mathematics Education* (pp. 165-170). London: PME.

Shigematsu K., Katsumi Y., Katsui H. and Ikoma Y. (2002). A research on metacognitive support for students by using the method of math journal writing for middle grades elementary mathematics. *Arithmetics Education (Journal of Japan Society of Mathematical Education)*, 84 (4), 10 -18. (in Japanese)

CHAPTER 63

CROSS-CULTURAL STUDIES ON MATHEMATICS CLASSROOM PRACTICES

Yoshinori Shimizu
University of Tsukuba

EXTENDED ABSTRACT

The findings of large-scale international studies of classroom practices in mathematics include aspects of instruction as identified with a resemblance among participating countries while instruction in Japan seemingly unique. Japanese mathematics teachers, for example, appeared to spend more time on the same task in one lesson than their counterparts in other countries by having students work on a challenging problem and discuss alternative solutions to it. Also, experienced teachers in Japan typically highlighted and summarized the main points of lessons mostly at the final phases of entire lessons to have their students reflect on what they have learned. Started with these striking characteristics of mathematics classroom instruction in Japan, the author reviews and discusses findings of cross-cultural studies on mathematics classroom practices.

The video component of the Third International Mathematics and Science Study (TIMSS) was the first attempt ever made to collect and analyse videotapes from the classrooms of national probability samples of teacher at work (Stigler & Hiebert, 1999). Focusing on the actions of teachers, it has provided a rich source of information regarding what goes on inside eighth-grade mathematics classes in Germany, Japan and the United States with certain contrasts among three countries. One of the sharp contrasts between the lessons in Japan and those in the other two countries relates to how lessons were structured and delivered by the teacher. The structure of Japanese lessons was characterized as "structured problem solving".

With a brief review of the literature on comparative study of mathematics classroom practice, including the Learner's Perspective Study (Clarke, Keitel & Shimizu, 2006), the author discusses that the findings of large-scale international studies of classroom practices in mathematics suggest that "structured problem solving" in the classroom with an emphasize students' alternative solutions to the problem can be a characterization of Japanese classroom instruction from a teacher's perspective. Also, a coherence of the entire lesson composed of several segments, students' involvement in each part of the lesson, and the reflection on what they did are all to be noted for the quality instruction in Japanese classrooms.

Associated with the descriptions of "structured problem solving" approach to mathematics instruction discussed above, several key pedagogical terms are shared by Japanese teachers. These terms reflect what Japanese teacher value in planning and implementing lesson within Japanese culture. "Hatsumon", for example, means asking a key question to provoke and facilitate students' thinking at a particular point of the lesson. The teacher may ask a question for probing students' understanding of the topic at the beginning of the lesson or for facilitating students' thinking on the specific aspect of the problem. "Yamaba", on the other hand, means a highlight or climax of a lesson. Japanese teachers think that any lesson should include at least one "Yamaba". This climax usually appears as a highlight during the whole-class discussion. The point here is that all the activities, or some variations of them, constitute a coherent system called as a lesson that hopefully include a climax. Further, among Japanese teachers, a lesson is often regarded as a drama, which has a beginning, leads to a climax, and then invites a conclusion. The idea of "KI-SHO-TEN-KETSU" which was originated in the Chinese poem, is often referred by Japanese teachers in their planning and implementation of a lesson. It is suggested that Japanese lessons has a particular structure of a flow moving from the beginning ("KI", a starting point) toward the end ("KETSU", summary of the whole story).

For understanding what Japanese teachers value in their instruction with cultural influence on them, a story or a drama can be a metaphor for characterizing an excellent lesson. Comparing classroom practice across cultures allows us to examine our own teaching and learning in classroom from a new perspective by widening the known possibilities. In addition to examining how teachers and learners across one's own country approach mathematics, opening up the lens to include an examination of how teachers and learners in other cultures approach the same topic can make one's own teaching and learning practices more visible by contrast and therefore more open for reflection and improvement. Comparing classroom practice across cultures can reveal alternatives and stimulate discussion about the choices being made within a country.

Keywords: Japanese classroom instruction, cross-cultural studies, Learners' Perspective Study, mathematics teaching

REFERENCES

Clarke, D., Keitel, C. & Shimizu, Y. (2006). The Learners' Perspective Study. In D. Clarke, C. Keitel &Y. Shimizu (eds.) *Mathematics Classrooms in Twelve Countries: The Insider's Perspective*. Rotterdam: Sense Publishers.

Stigler, J. W. & Hiebert, J. (1999). *The Teaching Gap: Best Ideas from the World's Teachers for Improving Education in the Classroom*. New York: NY, The Free Press.

CHAPTER 64

SYSTEMATIC SUPPORT OF LIFE-LONG PROFESSIONAL DEVELOPMENT FOR TEACHERS THROUGH LESSON STUDY

Akihiko Takahashi
DePaul University

EXTENDED ABSTRACT

For more than a decade, educators and researchers in the field of mathematics education have been interested in lesson study as a promising source of ideas for improving education. Takahashi discusses the research on the Japanese approach of professional development for mathematics teachers with the focus of lesson study by reviewing the research conducted by not only Japanese researchers but also by U.S. researchers who have carefully looked at lesson study.

An enormous interest in lesson study from other countries in the late 1990s made Japanese mathematics education researchers realize that lesson study is a unique approach to teacher professional development in mathematics education. At the same time, many Japanese mathematics education researchers have come to understand the need for in-depth research on lesson study in mathematics. For example, even U.S. researchers have pointed out that Japanese mathematics classrooms have changed over time by implementing ideas from research using lesson study (C. Lewis & Tsuchida, 1998; Stigler & Hiebert, 1999; Yoshida, 1999b). How lesson study actually contributes to the changes in teaching and in student learning are still unclear. Some other U.S. researchers have asked Japanese mathematics education researchers what the definition of lesson study is, but Japanese researchers have had a hard time clearly defining it because lesson study can be seen in many different settings, and in different forms and scales.

As a result, several studies about the different forms and scales of lesson study in mathematics were conducted and published during the 2000s. Using some of these studies, the author discussed the different purposes and forms of lesson study to describe a framework of Japanese mathematics teacher professional development. This includes pre-service and in-service as well as the role of lesson study plays in its various forms and purposes.

Although Japanese teachers and educators have been deeply involved in lesson study for more than a century, this might be the first time for educators and researchers around the world to become interested in learning about the Japanese professional development approach. Makinae (2010) argues that the origin of Japanese lesson study, *jyugyo kenkyu*, was influenced by U.S. books written for educators to introduce new approaches in their teaching during the late 1880s. He pointed out that one book by Sheldon (1862) describes methods to learn about new teaching approaches known as, *criticism lesson* and *model lesson*, which may be the beginning of Japanese lesson study. In fact, Inagaki (1995) argues that the *criticism lesson* was already practiced among elementary schools affiliated to the normal schools in Japan as early as the late 1890s. According to Makinae (2010), teacher conferences utilizing *criticism lessons* were conducted by local school districts in the early 1900s. Some of these conferences were already called lesson study conferences, *Jugyo-kenkyu-kai* in Japanese.

Considering the origin of lesson study, *criticism lesson* and *model lesson*, and the two functions of lesson study, a variety of lesson study sessions practiced among schools in Japan can be found in-between the *criticism lesson* and *model lesson*. Several researchers have tried to sort out the forms of mathematics lesson study in Japan using several criteria resulting in 2 to 7 forms. These differences are based on how a researcher looks at each form of lesson study. Although the major part of the study conducted by the researchers focused on lesson study in in-service professional development, lesson study also plays an important role in pre-service education in Japan. During teacher preparation programs, Japanese pre-service students in teacher training institutions usually have both opportunities: observing cooperating teachers' teaching, and teaching lessons under the guidance of cooperating teachers. Using the experiences during student teaching, the pre-service students are expected to become life-long learners. Lesson study plays an important role for Japanese teachers to continuously learn throughout their career, which sometimes is as long as 40 years, to improve teaching by working with colleagues. One reason that Japanese mathematics educators and teachers expect other teachers to be life long learners is because they share the belief that it is impossible for teachers to become effective by just completing teacher preparation programs.

Japanese mathematics educators and teachers identify three levels of expertise to teach mathematics; the teachers who can teach important basic ideas of mathematics such as facts, concepts, and procedures (Level 1), the teachers who can explain meanings and reasons for important basic ideas of mathematics in order for students to understand them (Level 2), and teachers who can provide students opportunities to understand these basic ideas, and support their learning in order for students to become independent learners (Level 3) (Sugiyama, 2008). Japanese mathematics educators and teachers share the belief that it is essential for teachers to continuously learn because by just going through teacher preparation programs may not be enough for a majority of teachers to develop knowledge and expertise for becoming level 3. The readers will see how Japanese universities and school districts share the responsibility in supporting pre-service teachers to become life-long learners using lesson study.

Possible future research related to lesson study would be the study of the implementation of not only lesson study but also a professional development structure like the Japanese one in different cultures. This could bring about new insights for researches working with various school systems in the world.

Keywords: Japanese lesson study, trajectories of teacher development, mathematics teacher professional development

REFERENCES

Inagaki, T. (1995). *A Historical Research on Teaching Theory in Meiji-Era.* Tokyo: Hyuuron-Sya.
Lewis, C., & Tsuchida, I. (1998). A lesson like a swiftly flowing river: Research lessons and the improvement of Japanese education. *American Educator, 22*(4).
Makinae, N. (2010). *The Origin of Lesson Study in Japan. Paper presented at the The 5th East Asia Regional Conference on Mathematics Education: In Search of Excellence in Mathematics Education,* Tokyo.
Sheldon, E. A. (1862). *Object-teaching.* New York: Charles Scribner.
Sugiyama, Y. (2008). *Introduction to elementary mathematics education.* Tokyo: Toyokan.

INDIA

SECTION EDITOR
K. Subramaniam
Homi Bhabha Centre for Science Education,
Tata Institute of Fundamental Research

Editorial Board:

Amitabha Mukherjee, *University of Delhi*
Farida A. Khan, *Jamia Millia Islamia*
R. Ramanujam, *Institute of Mathematical Sciences C.I.T. Campus, Taramani*
Rakhi Banerjee, *Ambedkar University, Delhi*

CHAPTER 65

EVOLVING CONCERNS AROUND MATHEMATICS AS A SCHOOL DISCIPLINE

Curricular Vision, Classroom Practice and the National Curriculum Framework (2005)

Farida Abdulla Khan
Jamia Millia Islamia, New Delhi

EXTENDED ABSTRACT

The history of schooling in independent India, the developmental trajectory that was adopted by a newly independent Indian nation, and the ensuing national and educational policies have all contributed to an increasingly content-loaded curriculum for mathematics at the school level. The subject forms an integral and highly valued component of the Indian school curriculum and is compulsory up to class 10, which signals the end of the common and compulsory school curriculum and is marked by an important qualifying exam at the national level.

The policy rhetoric of independent India has repeatedly endorsed education as a major instrument of social change, primarily through its ability to inculcate scientific rationality and modernization and to enhance 'development' through scientific and technical education. As early as 1952, the Secondary Education Commission, commenting on the challenges to education and the problems facing the nation, makes reference to "one of its most urgent problems – if not the most urgent problem – is to improve productive efficiency, to increase the national wealth and thereby to raise appreciably the standard of living of the people". Mathematics, within the policy perspective, was to establish itself as a foundational discipline that was to carry the burden of leading the country towards the developmental goals – social and economic – that were being envisaged.

The two major policy initiatives on school education in independent India – the Education Commission of 1964-66 and the National Policy on Education 1986/1992 – gave a central status to the development of Science and Technology and underlined the critical role of mathematics in achieving this objective. Both documents refer to mathematics as a foundational discipline and emphasize its contribution to the development of science, science research and technology. Underlining the importance of mathematics, the Education Commission, states that "One of the outstanding characteristics of scientific culture is quantification. Mathematics therefore assumes a prominent position in modern education. Apart from its role in the growth of the physical sciences, it is now playing an increasingly important part in the development of the biological sciences. The advent of automation and cybernetics in this century marks the beginning of the new scientific industrial revolution and makes it all the more imperative to devote special attention to the study of mathematics. Proper foundation for the knowledge of the subject should be laid at school" (Education Commission, 1966). In the present context of economic globalization, advances in science and technology and the domination of multinational business corporations, knowledge and expertise of mathematics is seen as a critical asset. As a school subject it becomes indispensable for all manner of technological, scientific, economics and business courses at the level of higher education and enables access to a variety of prestigious and highly valued occupations.

As curriculum restructuring in the aftermath of these initiatives tended more and more towards keeping pace with the increasing knowledge base of mathematics, students, teachers and many other voices from within and outside the educational establishment began to express their concerns regarding the school curriculum, its disconnect from lived experience and the constraints to its transaction in any meaningful or creative way in the classroom. The Yashpal Committee report entitled "Learning without Burden" which appeared in 1992 was a response to the extremely stressful learning routines and an equally stressful examination system that was becoming a defining feature of Indian schooling. It was a well grounded critique that helped to highlight some major problems of school education and touched a chord in the general population as well as the educational establishment. This document was a major inspiration for an intensive exercise of revising the entire school curriculum in India that the National Council for Educational Research and Training undertook, to produce the National Curriculum Framework (NCF) in 2005.

This chapter analyses the mathematics component of the NCF 2005 as the latest initiative in the history of educational and curricular restructuring and reform in India since independence. While retaining its commitment to the objectives and ideals of the Indian constitution and the Education Commission of 1968, this curricular framework sets itself up to counter the teaching-learning practices that have dominated Indian classrooms and alienated the school child. The NCF 2005, its manifestation in the form of a revised syllabus and a radically revised set of textbooks is premised on a social constructivist theory of knowledge and what it terms as critical pedagogy. The primacy of the child as an active learner is strongly emphasized and the child's psychological and social development are central concerns. While presenting a critique of contemporary Indian society and the implications this has for education, it seeks to challenge privileged knowledge systems and the rigid hierarchies of schooling practice. This chapter traces this shift in perspective and discusses its salient features especially with reference to the mathematics curriculum. The classroom however, is not a decontextualised space but functions within larger social, political and economic structures and an educational system that mirrors all manner of social injustice, economic hierarchies and immense disparities. Reflecting on the mathematical component of the new curriculum, the chapter comments upon the resistance that any challenge to an established system of privileged access is likely to encounter, and the implications this has within the classroom and beyond it.

Keywords: policy, history of mathematics education, curriculum policy, educational systems in India, India

REFERENCES

Government of India. 1956. *Report of the Secondary Education Commission.* Ministry of Education, New Delhi: 3rd reprint, 1956. Hind Union Press, New Delhi. (Publication No. 165)

Government of India. 1966. *Report of the Education Commission (1964-66): Education and National Development.* Ministry of Education, New Delhi.

CHAPTER 66

CURRICULUM DEVELOPMENT IN PRIMARY MATHEMATICS

The School Mathematics Project

Amitabha Mukherjee
University of Delhi

Vijaya S. Varma
Ambedkar University, Delhi

EXTENDED ABSTRACT

The School Mathematics Project (SMP) was an experiment in primary mathematics curriculum development and a small-scale intervention in the Indian school system carried out over 8 years from 1992-2000. It is an important link in a chain of programmes of curricular reforms and material development in elementary schools in India that includes the Hoshangabad Science Teaching Programme and a number of smaller programmes. SMP started in 1992, initially as a discussion group based at the Centre for Science Education and Communication (CSEC), University of Delhi. The stated aim of the group was to try and develop a programme that would address the fear of mathematics in school children. Initial exploratory meetings led to the formation of a core group that included school teachers, university teachers, researchers and others interested in the teaching of mathematics in schools. From April 1993, the project was supported by a grant from the Ministry of Human Resource Development, Government of India.

The first phase of activities consisted of a series of discussion meetings, a national conference and three studies. The conference, held in October 1993, was on "The Basis for Curricular Choices," and was attended by a number of participants from all over the country. It was an attempt to address a very real problem the group was then facing – what basis to use in deciding what should or should not be included in a curriculum in school mathematics, particularly the early years. The first study was on teachers' attitudes towards children and mathematics, while the second was a street survey on the extent ordinary people use mathematics in daily life to see whether or not this could be used as a basis for designing the school curriculum in mathematics. The third study was a baseline survey to ascertain the mathematical readiness of students who had just been admitted to Class 1 (Grade 1). These initial interactions and studies convinced the core group that any intervention in school mathematics had to begin from Class 1.

In the second phase, there was interaction with a larger group in Delhi. To identify the intervention schools and initiate a dialogue with teachers, a series of meetings called *Ganit par Gupshup* (Chatting about Mathematics) was started. These went on over a period of several months. In the winter of 1994-1995, three material creation workshops were organised. The aim was to create materials that were consistent with the group's approach and could be used in the classroom in Class 1 in the intervention schools.

During this period, and in parallel with the above activities, the group was in the process of articulating its approach. In March 1995 a formal document entitled "The SMP Approach" was ready. Some of the key points of the approach are as follows:

- Children are not blank slates when they enter school. We have to take cognizance of the initial mathematics they bring with them to school.
- Children are individuals with their own pace of learning, and there should be room for them to remain different.
- Handling concrete objects must form an important component of early work in mathematics before any level of abstraction is introduced.
- The progression should be Concrete → Oral-Contextual → Abstract.
- Mathematics is more than number, operations and algorithms. It encompasses shape and space, patterns, structures, data handling and measurement.
- Mathematics is inherently beautiful and a potential source of joy – but only if the teacher feels this herself can she communicate it to children.
- Facility comes naturally when there is a meaningful context for mathematics.

The third phase, starting in April 1995, consisted of intervention in five schools, representing a mix of school types. Two of these were regular Government schools (Sarvodaya Schools), while one was a school managed by the University of Delhi. The remaining two schools belonged to the private category, although they were a special kind of school managed by the armed forces.

Teachers who were to teach Class 1 in the five schools administered the baseline survey on their own students who had just been admitted in Class 1. Subsequently, they went through an intensive orientation programme before embarking on teaching according to the new curriculum. The output of the material creation workshops had been edited and compiled into a volume entitled *Games and Activities for Class I*, which, in draft form, served as a manual for the teachers. Over the rest of the academic year the programme unfolded in the target schools. Classroom teaching was supplemented by regular feedback meetings, which were held by turns in the five schools, and by follow-up school visits by core group members. In two of the schools, regular classroom observation was carried out by a research associate, who maintained a detailed diary. At the end of the academic session, the *Games and Activities* volume was published in book form by CSEC.

In March 1996, when the programme was about to complete the first year of work in the classroom, financial support was abruptly withdrawn by Ministry of Human Resource Development. This step did not appear to be based on any assessment of the work carried out under the project, and the reasons were never communicated to the group. At an emergency meeting, the programme teachers expressed their resolve to carry on. The programme thus continued with skeletal support from the core funds of CSEC. However, the absence of funds did have an impact on the programme. Photocopying of materials had to be restricted. The feedback meetings became irregular, as did school visits. Classroom observation had to be abandoned altogether, since the salary of the researcher could not be paid.

For Class 2, while a lot of the material in the Class I book could be used directly, teachers created other materials as and when needed. Although feedback meetings were no longer regular, they were still frequent enough at this point to think of the material as being evolved collectively. At some point during this year, one of the Sarvodaya Schools dropped out of the programme.

During the years 1997-1999, the intervention moved to Classes 3 and 4. Meetings were held largely on a felt-need basis. However, while materials used in the classroom were pooled together for Class 3, this was never done for Classes 4 and 5. Thus the classroom transaction was based completely on the individual teachers' interpretation of the group's approach. During this time, the second Sarvodaya School also dropped out of the programme, while the participation of the Delhi University school became marginal.

In January 1999, CSEC organised and hosted a Seminar on 'Aspects of the Teaching-learning of Primary Mathematics'. This was attended among others by participants from a variety of leading academic institutions and Non Governmental Organisations from across the country. A number of papers were presented, based on the wide experience of participants. The SMP group also presented its concerns and ongoing work. It is worth noting that the presentations were not only made by members of the original core group, but also by some of the teachers in programme schools, reflecting their growing confidence.

In 1999-2000 the programme got some financial support, as it came under the ambit of CSEC's project Elementary Education Teachers' Research Network (EETRN) – a collaborative programme between Homerton College of the University of Cambridge and CSEC. This led to friutful exchanges and sharing of experiences of SMP teachers with teachers working in very different settings across the country.

In April 2000, the children who had been in the first intervention batch of SMP entered middle school (Class 6). An evaluation study of the five years of the intervention was carried out later that year. This consisted of interviews with teachers and children as well as a written test administered to children in two programme schools and to a comparable group of children in a non-programme school. Some of the key findings include an increase in the self-confidence of the teachers and a greater ability on their part to articulate their understanding of how mathematics should be taught in the lower classes. The interviews with children showed that they were not afraid of mathematics. However, the written test, carried out after more than six months of conventional instruction in the sixth Class, did not show a significant difference between the SMP children and the children in the control group.

According to some teachers, SMP has never been closed down, and continues to be a live programme in their classroom practice. However, formally, no specific activities were undertaken after the evaluation study, except that the Class 2 materials were edited and brought out in 2002 in the form of a book for teachers and a workbook.

The SMP experience of over eight years has taught people directly involved with it many things which would be hard to summarise. Nevertheless, one can abstract some lessons learned from it which would be useful for any future programme of intervention in school mathematics. The first point is that it is indeed possible to create and implement a curriculum for primary school mathematics that leads to freedom from fear. However, to keep it going needs constant work, and teachers need all the support they can get. In this, the role of the core group is crucial. The second lesson is that it is possible for teachers to not only teach in a different way, but to create new materials for their own classroom use and to reflect on and critique their own practice. This assumes, of course, appropriate orientation and continuing support. The third point which emerges is that we need to work with a cross-section of mainstream schools. However, we have to acknowledge that the government schools are hard to retain, and need special attention. This, of course, is not specific to mathematics.

As an epilogue, we may note that SMP appears to have had an influence that goes beyond its small scale and its difficulties on the ground. Some SMP ideas have been absorbed into mainstream curriculum design in India. Materials prepared by the Indira Gandhi National Open University show the influence of SMP. The textbooks developed by CSEC for the Delhi State

Council of Educational Research and Training reflect the strong influence of SMP in the emphasis given to specific areas as well as in the arrangement of materials. The Delhi textbook effort played a role in the National Curriculum Framework (NCERT 2006a) of NCERT. Generally, the SMP emphasis that non-number areas of mathematics – Shapes and Space, Patterns, Measurement and Data Handling – should get their due place in the primary classes finds an echo in the Position Paper of the National Focus Group on the Teaching of Mathematics. In particular, the present authors had advocated in the 1999 Seminar that operations on fractions be postponed to higher classes, and the 2005 NCERT syllabus has done precisely that (NCERT 2006b). The current primary mathematics textbooks of NCERT also show the imprint of SMP (NCERT 2006c; 2007; 2008).

Keywords: curriculum development, India, SMP, mathematics syllabus

REFERENCES

NCERT (2006a). *Position Paper: National Focus Group on the Teaching of Mathematics.* New Delhi: National Council of Educational Research and Training. Retrieved from http://www.ncert.nic.in/new_ncert/ncert/rightside/links/pdf/focus_group/math.pdf

NCERT (2006b). *Syllabus Volume I: Elementary level.* New Delhi: National Council of Educational Research and Training.

NCERT (2006c, 2007, 2008). *Math-Magic 2, 3, 4, 5.* New Delhi: National Council of Educational Research and Training.

CHAPTER 67

INTERVENING FOR NUMBER SENSE IN PRIMARY MATHEMATICS

Usha Menon
Jodo Gyan, Delhi

EXTENDED ABSTRACT

The paper reports on innovations made in the use of the empty number line while intervening for supporting early number sense in primary mathematics in India. The innovations were made within a non-profit social enterprise called Jodo Gyan and emerged from what can be characterised as 'research informed practice'. This set of innovations is linked to the relationship between the quantity and order aspects of number.

Keywords: number sense, number line, word problems, classroom didactics, primary curriculum in India

Empty Number Line

The Empty Number Line (ENL) that emerged from the school of Realistic Mathematics Education (RME) in the nineties is a major contribution for supporting the development of number sense with flexibility to deal with addition and subtraction. The earlier attempts at incorporation of the (numbered) number line into the classroom was ineffective since it called forth counting and reading-off responses from children and not higher order structuring responses (Gravemeijer, 2002, p. 11).

The situation changed when following an idea of the American mathematician Hassler Whitney, Treffers introduced the ENL (Treffers & de Moor, 1990). The use of ENL was chosen because it provided a tool to do addition and subtraction using the 'stringing' method which not only connected with the forward counting strategy that children naturally use to do addition (Klein, Beishuizen & Treffers, 1998) but also was seen to avoid the pitfalls associated with the splitting strategy, especially in subtraction (Beishuizen, 1993).

The original formulation of the ENL by Hassler Whitney as well as by Adri Treffers had linked it to the use of a ten structured bead string on which activities such as location of numbers were done by means of a clip put after the last counted bead.

ENL With or Without the Bead String

In the extension of the use of the ENL to many other countries such as U.K., Australia and New Zealand in the last decade, the use of the bead string appears to be getting a lesser or even no role. The use of the ENL has been extensively propagated in England, especially since 1999, with the introduction of the National Numeracy Strategy. A perusal of the wide range of the documents produced indicates a much stronger role for the numbered number line as a prelude to introducing the ENL rather than for the bead string. The U.K. Department for Education document entitled 'Teaching children to calculate mentally' presents details of the ENL, but hardly anything on the activities to be done with the bead string. Further the nature of the presentation of the bead string (Department for Education, 2010, pp. 18-19, 29), seems to indicate that the ENL is being seen as a static 'model of the bead string' rather than as a model of 'working with the bead string' (Gravemeijer, 2002, p. 14).

Bobis (2007, p. 12) from Australia informs us that the ENL was not introduced through a bead string in her daughter's class but with a marked number line. Unfortunately the strength of the ENL to support mental calculation through its flexibility seems to be getting compromised in the Netherlands itself where structured work on the marked number line is incorporated in many textbooks (van den Heuvel-Panhuizen, 2008).

The numbered number line in fact supports the ordinal aspect of number and not the quantitative or cardinal aspect. Using it to do addition and subtraction can give rise to the familiar mistake that children make of counting on from the first number – in adding 5 and 3, counting 3 further from 5 rather than from 6 and therefore reaching 7 instead of 8. There are also differences in the cognitive processes involved in adding 3 and 5 mentally or by counting on the numbered number line (Thompson, 2003). Therefore the numbered number line cannot be considered as a prelude to the ENL. Early experience with the introduction of the ENL without being preceded by activities with the bead string by Gravemeijer showed that children were unable to decide whether showing jumps from 90 to 88 on the ENL represented a take away of 2 or of 3 (Gravemeijer, 2002, p. 17).

Since 2000, I have been exploring the use of the ENL while working with children in Delhi and the trajectory as it evolved was shared with resource persons in Jodo Gyan. In this trajectory, the cardinal aspect has in fact got further strengthened through the introduction of specially designed activities with the bead string or 'Ganit Mala' as it is called by Jodo Gyan in India. Activities such as Clip 1 and counting by the 'position of the number' on the bead string after the introduction of zero are changes that have been introduced.

Supporting Ordinal/Cardinal Switch

The relationship between the ordinal and cardinal aspects in the use of the Ganit Mala can be understood by studying the process of locating numbers on the bead string. When beads are just counted the ordinal aspect of number is foregrounded. When a clip is put after counting that many beads to locate a number, the cardinality aspect emerges where the clip represents the position where there are that many beads from the beginning till that point. Through the specially designed activity of Clip 1, this cardinal aspect is underlined in the Jodo Gyan trajectory, since children are asked 'how many beads from the beginning till the clip?' After the introduction of the position for zero, counting takes place on the Ganit Mala by counting by the position first from 0 to 100 and then from 100 to 0. The changed practice of counting by the position facilitates reverse counting to locate numbers which otherwise would not be possible. (If children would count the beads from the end as 100, 99, 98, 97 and put the clip to locate 97 after the last counted bead, then the position would be that of 96). Thus counting on the Ganit Mala evolves from counting by the beads to counting by the position.

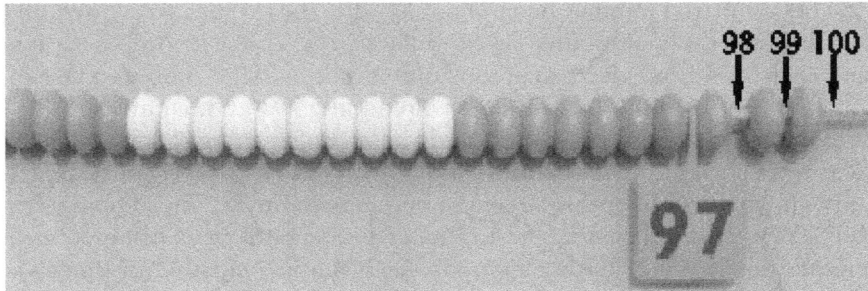

Figure 67.1.   Counting by the position.

Through this process of locating numbers through forward and backward counting using the ten-structure and through activities such as skip counting children develop an internal image of the number line, which then supports them to do addition and subtraction mentally with ENL. In the trajectory that we follow the children do location on the ENL after having done both forward and backward counting. It is normally seen that children are able to do the location easily once they are able to locate on the Ganit Mala in both ways – although in this case only the order aspect is strictly taken into consideration and not the linearity. Recent studies by Siegler and colleagues have shown the significance of similar (linear) activities for supporting number sense (Siegler & Ramani, 2009).

The paper presents the detailed trajectory of activities being followed by Jodo Gyan and compares it with the original Treffers trajectory.

Glimpses From the Classroom

Flexibility in the use of operations as per the number combinations, rather than carrying out a fixed computing procedure after finding out whether to add or subtract, can be considered as an indicator of number sense. This aspect is explored in the case of children following the Jodo Gyan curriculum in Grade II from two schools, in the case of the following word problem which needs subtraction of 85 from 92.

1. Sohail is reading a book of 92 pages. He has read 85 pages. How many pages does he need to read to complete the book?

Subtraction can be broadly done by two different strategies – by subtraction [either by subtracting the number (SUB) or by 'taking away to' the number (TAT)] or by 'adding on to' the number (AOT). 58 % of the children used the AOT strategy for this problem involving the pages to be read. While these same children used subtraction as a strategy to solve the word problem given below, indicating a certain flexibility.

The paper also presents the analysis of the procedures used by the children for the following question.

1. At the Sultanpur National park there were 74 white storks in the beginning of February. As it started getting hot they started leaving the park. 13 white storks left the park in the beginning of March and 22 left at the end of March. How many white storks were still there in the park?

Children used two strategies to solve the problem. Some children subtracted both the numbers one after another while the others chose to first add the number of birds that had left and then subtract it from the original 74 birds. In both these cases there were children who subtracted at one-go while others split the numbers as per their convenience and subtracted it.

There was wide variability in the procedures used to solve this problem even in a single class. In the different sections of class, with groups of children ranging from 18 to 27, there were on an average 2.1 children per procedure, with many classes having an average of 1.6, showing that in many cases only one child was using a particular procedure. This is a very different situation from what happens when the standard algorithm is taught where all the children who solve the problem correctly use the same procedure. This differentiation in procedures can be considered as an indicator of the number sense existing in the class where addition and subtraction are done on the ENL, following the Jodo Gyan trajectory.

Interacting With the School Community

We have found that the didactical context has a crucial significance while sharing the teaching-learning experience with the teachers. Reflection and interaction with the teachers over the years shows that the teachers have included the trajectory to varying levels in their own teaching. This could also be because doing addition and subtraction on the ENL is unfamiliar to most of the teachers since almost all the schools use a place-value based vertical algorithm for teaching addition and subtraction. Accordingly, the interactions with teachers have as targets different levels to which teachers may incorporate the learning trajectories as below:

- Sharing the complete trajectory including the use of the ENL for doing addition and subtraction – done through a 2 year programme involving biweekly visits tapering off in the later years and stabilizing at half yearly assessments and workshops.
- Sharing all the Ganitmala activities and only the activity of locating numbers on ENL – usually done during 3-10 day workshops on primary mathematics.
- Sharing only the activities that involve forward jumps such as Clip 1, Clip 2 and so on including one number to another and no marking on the ENL – shared during introductory workshops of one day or less.

Conclusion

The experience of intervening for number sense indicates the continuing importance of building on the quantitative intuitions of children. In the case of early numbers, the use of the bead string or Ganit Mala supports the interlinking of the cardinal and ordinal aspects and indicates that extended activities with the bead string can support the development of number sense.

REFERENCES

Beishuizen, M. (1993). Mental startegies and materials or models for addition and subtraction up to 100 in Dutch second grades. *Journal for Research in Mathematics Education* (pp. 294-323).

Bobis, J. (2007). ENL: A useful tool or just another procedure? *Teaching Children Mathematics* (pp. 410-413).

Department for Education, U.K. (2010). *Teaching Children to Calculate Mentally.* Retrieved from http://dera.ioe.ac.uk/778/1/735bbb0036bed2dcdb32de11c7435b55.pdf

Gravemeijer, K. (2002). Didactisch gebruik van de lege getallenlijn – een persoonlijk perpsectief. (Didactical use of the empty numberline – a personal perspective). *Panama-Post, Tijschrift voor nascholing en onderzoek van het rekenwiskunde onderwijs* (pp. 11-23).

Klein, A. S., Beishuizen, M., & Treffers, A. (1998). The empty Number Line in Dutch Second Grades: Realistic versus Gradual Program Design. *Journal for Research in Mathematics Education* (pp. 443-464).

Siegler, R.S. & Ramani, G.B. (2009). Playing Linear number board Games – But not Circular Ones- Improves Low-Income Preschooler' Numerical Understanding. *Journal of Educational Psychology* (pp. 545-560).

Thompson, I. (2003). On the right track?. *Primary Mathematics*, Autumn (pp. 9-11)

Treffers, A., & de Moor, E. (1990). *Prove van een nationaal programma voor het rekenwiskunde onderwijs op de basisschool Deel 2 (Specimen of a National Programme for Primary Mathematics Education]*. Tilburg: Zwijsen.

van den Heuvel-Panhuizen, M. (2008). Learning from Didactikids: An Impetus for Revisiting the Empty Number Line. *Mathematics Education Research Journal* (pp. 6-31).

CHAPTER 68

SOME ETHICAL CONCERNS IN DESIGNING UPPER PRIMARY MATHEMATICS CURRICULUM

A report From the Field

Jayasree Subramanian and Sunil Verma
Eklavya, Madhya Pradesh

Mohd. Umar
ICICI Foundation for Inclusive Growth, Rajasthan

EXTENDED ABSTRACT

Curriculum development and textbook making in India are done at the national level by the National Council for Educational Research and Training (NCERT) and also at the individual state levels by the respective state councils. There is usually a large overlap in the content though there would be some difference in the approaches adopted by the different boards. While it is not mandatory for the schools to follow the textbooks brought out by the state, a large percentage of schools follow the state curriculum and the textbooks. However, schools in the urban and semi urban localities may have access to alternative textbooks, supplementary reading material for teachers and teaching aids brought out by commercial houses as well as by grass root level Non-governmental Organisations (NGO) trying to address the quality and equity issues together.

The mathematics curriculum research and material development team at Eklavya, an NGO with many years of engagement with education, is presently attempting to develop alternative curricular material in the form of modules at the upper primary level. The main purpose behind our efforts in producing alternative materials is that they function as supplementary reading for teachers and resource material for teacher development, though it is also hoped that this would inform the curriculum development work at the state and central level. Therefore the alternatives focus on themes that are part of the prescribed curriculum at the upper primary level and are informed by current research and classroom trials.

Children at the upper primary level are expected to move beyond everyday mathematics and engage with abstraction, acquire logical thinking, ease with symbolic representation and competence to do mathematics at the higher classes (National CurriculumFramework 2005). (Though the term "upper primary" is largely used to refer to Grades 6 to 8, it is also sometimes used to refer to Grades 5 to 8 or 5 to 7.) Typically the upper primary curriculum deals with rational numbers and their properties, algebra, geometry, data handling and 'commercial mathematics'. This article would discuss the context, namely government run schools in a semi-urban locality and private schools catering to the lower income group in urban and semi-urban localities within which explorations were made and the design experiments were conducted, the socio-economic background of the children, their competence at the upper primary level and the implication of what is taught at the upper primary level for these children. Following the critical approach to mathematics education, the article would argue that curriculum designing involves ethical issues and that the curricular choices made by the boards of education, rather than representing the needs and interests of these children, function to further marginalize them. It would also stress the need for an alternate vision of upper primary mathematics curriculum that serves the interests of the majority of the children rather than the interest of the discipline alone.

Curriculum Development in Upper Primary Mathematics

Systematic curricular research and material development in mathematics education is still an emerging area in India. This is partly because education as an academic domain has not received the kind of attention it deserves from the state. Education departments in the country have largely focused on developing and implementing pre-service teacher education programs. In fact barring a couple of institutions that are engaged in education research and curricular research, most of the experimentation and innovation in education have come from the NGOs which often function to address the needs of the marginalized sections and have to work against a range of constraints. Research publication from the country therefore does not match the efforts that have gone into intervention in education.

At the upper primary level, the Homi Bhabha Center for Science Education (HBCSE) at Tata Institute of Fundamental Research and Eklavya have been actively engaged in design experiments in teaching and learning of algebra, geometry and rational numbers (fraction, ratio and proportions).

In Algebra the design experiment by the HBCSE team uses Grade 6 students' knowledge of arithmetic as a starting point to build a bridge between arithmetic and symbolic algebra. Using "terms" as objects that combine with other "terms", the experiment exposes children to a new way of looking at arithmetical expressions, operating on them and comparing them (Banerjee & Subramaniam, 2004; Banerjee & Subramaniam, 2005). This experience is then used in contexts such as generalizing, predicting, and proving, all of which require symbolic algebra. Literature on algebra suggests that children's understanding of algebra broadly follows the historical development of algebra in that children acquire procedural understanding initially and over a period of time acquire structural understanding. The "terms approach" enables children to acquire structural understanding alongside procedural understanding by requiring children to initially decompose arithmetic expressions and later with that experience algebraic expressions into signed "product terms" which are first evaluated (in the arithmetic context) and then "combined" (meaning added) (Banerjee, Subramaniam & Naik, 2008). In a recent paper Subramaniam and Banerjee argue that historically algebra has been seen as foundation to arithmetic in the Indian tradition and that the "terms approach" could be seen as drawing on that tradition at least to clarify ideas (Subramaniam & Banerjee, 2010). They use the term "operational composition" to denote information contained in an expression and argue that a refined understanding of operational composition includes judgments about relational and transformational aspects and it may play a role in developing understanding of functions and thus point to the importance of emphasizing the structural understanding of numerical expressions as a beginning for symbolic algebra.

In the teaching and learning of fractions a collaborative design experiment at HBCSE and Eklavya combines the share and measure interpretation of fractions to facilitate students' reasoning in making sense of the fraction symbol and in comparison tasks (Subramaniam & Naik, 2007; Naik & Subramaniam, 2008). In the Indian context fractions are introduced at Grade 3 level and the instruction uses either the part whole approach or measure approach. But soon the approach brings in algorithms for comparison of fractions and for introducing addition and subtractions of fractions. While some students may acquire skill in algorithmic computations many have difficulty with making sense of what they are doing. Alternative approaches developed by the two teams share a basic conviction that share and measure meaning of fractions enable children in acquiring a situated and conceptual understanding of the notion of fraction. Class room trials in four different sites demonstrate that children see the connection between the two and use the two approaches flexibly to reason in comparison tasks (Subramanian, Subramaniam, Naik, & Verma, 2008; Subramanian & Verma, 2009).

Curricular Explorations and Marginalized Children

A rare opportunity was provided by a state government in central India to the NGO Eklavya to develop and implement in some parts of the state an alternative curriculum. The curriculum called "Prashika," involved teachers in its process of development and was developed for predominantly first generation learners, many of them from economically disadvantaged and linguistically marginalized backgrounds. Prashika innovatively combines language and mathematics in the early years postulating a fundamental link between the two (Agnihotri, Khanna, & Shukla, 1994; Rampal, 2002). After the state government closed this program in the year 2000 our efforts have been directed towards providing outside school support at the primary school level to first generation learners in about 200 centers and developing alternative curricular material at the upper primary level.

The mathematics curricular research and material development team at Eklavya has been engaging with the upper primary mathematics curriculum with a view to understanding difficulties children face and exploring alternative approaches for teaching fractions, negative numbers, algebra and geometry for the last five years. Since children find rational numbers and its precursor at the primary school level "fraction" difficult, a considerable part of our effort focuses on developing an alternative approach for teaching fractions. Our initial trials began with teaching rational numbers at the Grade 7 level in a private school catering to children from the low income group in a city in central India and the classroom trials soon convinced us that we need to begin from the beginning. In collaboration with the HBCSE team we designed an alternative that combined the share and measure meanings of fractions and trialed it with two groups of children, one studying in a school run by an NGO for children working on daily wages and the other at a private school catering to lower income group in a small town. These trials convinced us that children make sense of fraction when they are introduced to it through a context that is meaningful to them and come up with their own arguments for solving problems. Arguments children used in tasks involving comparison, equivalence and the operation of addition are evidence that in their logical ability these children are no less than any other group of children (Subramanian et al., 2008). We also tried to teach negative numbers at Grade 6 and 7 level with limited success partly because even the best performing children had only an algorithmic comfort with numbers and some students – particularly girls – had difficulty writing numbers. However, when we moved to government run schools in a small town for classroom trials we found that a large number of Grade 6 children had difficulty reading and could not write three and four digit numbers and Grade 3 children had difficulty writing two digit numbers. While some of the Grade 6 children could do computations in their mind and solve problems, they could not write what they said. This meant that we really could not begin work on algebra and rational numbers without first teaching children numerical representation and operations on numbers. Our strategy therefore was to begin our work with Grade 3 for one set of children, focusing primarily on fractions but also improving their comfort with two digit numbers alongside and continue with the same set in the following years. Simultaneously we were also trying to work at least at a limited level with Grade 6 children in fractions, algebra and geometry. In geometry we focused on teaching angle measurement, and the use of protractor and we found that children could correctly estimate the measurement of the angle before actually measuring it. Any trial in algebra is very challenging because children cannot write simple arithmetic expressions or even numbers properly. In our effort to assess what level of abstraction children can engage with, we adopted a procedural approach and made children compute solutions for verbally stated arithmetic expressions and we wrote them down on the board and used them as basis to introduce algebraic expressions. While children participated in these exercises eagerly, it was clear that they would not be able to write these expressions on their own. The contradictions between children's potential, their willingness to participate in the classroom activities and the extreme marginalization they face in the schools raise some fundamental ethical challenges for curriculum designers.

Government schools are ideal sites for design experiments because a large majority of their students belong to the intersection of marginalization along the lines of class, caste, community and gender and many of them are first generation learners. While these children fail to attain minimum levels of learning in the mainstream school, there are studies that demonstrate the kind of informal mathematical knowledge that children from these backgrounds have and considerable evidence to show that they learn well in the alternative schools run by NGOs (Khan, 2004; Nawani, 2002). It is well argued that if a large number of children belonging to socio-economically disadvantaged sections remain untaught in spite of enrolling in school, a curriculum that aims to bring in the concerns of the discipline proper into school mathematics, serves only to ensure that these children dropout of school sooner or later (Skovsmose, 2005).

The article will discuss in detail the kind of marginalization that these children face in the school, discuss some of the causes behind it and raise some ethical considerations in designing curriculum at the upper primary level. The article would argue that rather than "the dumbing down approach" which forever denies opportunities for the marginalized, the upper primary curriculum needs to take cognizance of the strengths and limitations of all the children and evolve alternatives that result in meaningful learning experiences for them.

Implication and Suggestion for Future Work

Our work points to the need to study marginalization in a systematic way paying attention to various limiting factors such as caste, class, gender and the rural urban divide. There is also a serious need to revisit what we teach at the upper primary level and who are seen as disposable by the system. There is very little research that addresses the dilemmas of designing a curriculum that serves the interest of all the students and alternatives that make mathematics education worth the while for those belonging to the margins. We hope that our work would contribute to future explorations along these lines.

Keywords: ethics in curriculum development, design experiments in mathematics, curriculum in India, research and development in mathematics education

REFERENCES

Agnihotri, R., Khanna, A. L., & Shukla, S. (1994). *Prashika- Eklavya's innovative Experiment in Primary Education.* New Delhi:Ratna Sagar.

Banerjee, R., & Subramaniam, K. (2004). 'Term' as a bridge concept between arithmetic and algebra, *In Proceedings of Episteme-1*(pp. 76-77). Homi Bhabha Centre for Science Education, Mumbai.

Banerjee, R., & Subramaniam, K. (2005). Developing Procedure and Structure Sense of Arithmetic Expressions. In H. L. Chick & J. L. Vincent (Eds), *Proceedings of the 29th conference of the International Group of the Psychology of Mathematics Education*, Melbourne, Australia

Banerjee,R., Subramaniam, K., & Naik, S. (2008). Bridging arithmetic and algebra: Evolution of a teaching sequence. In O.Figueras et al. (Eds.) *International group of the psychology of mathematics education: Proceedings of the Joint Meeting of PME32 and PME-NA XXX (PME29)* (Vol 2, pp. 121-128). Morelia, Mexico.

Khan, F. A. (2004). Living, Learning and doing mathematics: A study of working class children in Delhi. *Contemporary Education Dialogue*, 1(2).

Naik, S., & Subramaniam, K. (2008). Integrating the measure and quotient interpretation of fractions. In O. Figueras et al. (Eds), *International group of the psychology of mathematics education: Proceedings of the Joint Meeting of PME 32 and PME-NA XXX (PME29)*, (Vol 4, pp.17-24). Morelia, Mexico.

Nawani, D. (2002). Role and Contribution of Non Governmental organizations in Basic Education. In R. Govinda (Ed.), Indian Education Report: A Profile of Basic Education. New Delhi, India: Oxford University Press

National Curriculum Framework. (2005). New Delhi: National Council for Education Research and Training. Retrieved from http://www.ncert.nic.in/html/pdf/schoolcurriculum/framework05/nf2005.pdf.

Rampal, A. (2002). Texts in Context: Development of curricula, textbooks, and teaching learning materials In R. Govinda (Ed.), *Indian Education Report: A Profile of Basic Education*. Oxford University Press

Skovsmose, O. (2005). Travelling through Education- Uncertainty mathematics Responsibility. Rotterdam: Sense Publishers.

Subramaniam, K., & Naik, S. (2007). Extending the Meaning of the Fraction Notation. *In Proceedings of Episteme-2* (pp. 223-227). Homi Bhabha Centre for Science Education, Mumbai

Subramaniam, K., & Banerjee, R. (2010). The Arithmetic-Algebra Connection: A historical-pedagogical perspective. *Advances in Mathematics Education*(pp.87-107). Dordrecht: Sringler

Subramanian, J., Subramaniam, K., Naik, S., & Verma, B. (2008). Combining Share and Measure Meaning of Fractions to Facilitate Students' Reasoning. *International Conference of Mathematics Education: ICME-11*. Monterrey, Mexico. Retrieved from http://tsg.icme11.org/document/get/823

Subramanian, J. & Verma, B. (2009). Introducing fractions using Share and Measure Interpretation: A report from Classroom Trials. *In Proceedings of Episteme-3*, Homi Bhabha Centre for Science Education, Mumbai.

CHAPTER 69

STUDENTS' UNDERSTANDING OF ALGEBRA AND CURRICULUM REFORM

Rakhi Banerjee
Ambedkar University, Delhi

EXTENDED ABSTRACT

Mathematics teaching-learning has been an area of concern for many teachers, parents, educators, and researchers since failure of students in the subject is a constant feature and mathematics is often a reason for dropping out of school. People do not largely question the importance of mathematics and justify its existence in the school curriculum on the basis of its 'supposed' utility in everyday life. For some the justification is that it is the gateway to access higher education in prestigious disciplines or professions.

Even though one tries to make some connection with students' daily lives in the primary grades, it is perhaps hard to do the same in the middle school level. The concepts/definitions and ideas grow in abstraction and the rules and the procedures gradually become delinked from concrete experiences, leading to a sense of arbitrariness in the symbol manipulation process. Studies in the past like, Strategies and Errors in School

Mathematics (Hart et al., 1981) and Concept in Secondary Mathematics and Science (Booth, 1984) have shown students' difficulties in understanding the symbols and procedures in various areas of mathematics. The few systematic studies done in the Indian context, indicate that the above findings are true also of Indian schools. (See Kanhere, Gupta & Shah, this volume.) It is essential for students to make the shift from mathematics based on concrete everyday experiences to one which is abstract, rule governed and symbolic in order to succeed in the lower secondary grades. A quick look at mathematics taught in the middle school level reflects this abstraction with emphasis on definitions and procedures and building the capacity to use conventional symbols.

In the recent times, mathematics only aimed at symbolic competence and ability to use rules and procedures in various problems (which leads to the exclusion of many from this strand of schooling) has been questioned. Like many other countries, India too responded to the challenge by engaging in a country wide deliberation resulting in a reformed curriculum document based on the constructivist philosophy, the National Curriculum Framework (NCF). NCF (2005) strongly recommends the need to move away from the narrow aim of "utility" of mathematics to focus on higher aims of developing "child's resources to think and reason mathematically, to pursue assumptions to their logical conclusion and to handle abstraction. It includes ways of doing things, and the ability and the attitude to formulate and solve problems" (p. 42). This surely requires the creation of a curriculum which teaches important/ meaningful mathematics, which is coherent and engages children in communicating with mathematics and posing and solving problems. Also, there is an underlying expectation of change in teacher attitude and action. They need to be well prepared to deal with the content and its pedagogic aspects as well as understand the motives of the change. The classroom interactions need to be redefined to be able to engage students in the "(re-)creation" of mathematics. Thus, curriculum reform has multiple dimensions and there are many determinants: textbooks, teachers, classroom, and assessment.

These deliberations and ideas led to the drafting of a new set of textbooks written by the National council for educational research and training (NCERT) that are being used by schools which are affiliated to the Central Board of Secondary Education since 2005. Compared to the older textbooks, the presentation in the new books is more lucid, addresses the students directly, there is an effort to lend meaning to the concepts and procedures and there is a change in the way the content is divided across grades. The assumption behind the change is to include more students in the teaching-learning process and achieve the 'higher aims' of mathematics learning in the elementary school. Since they are the major resource for most teachers (and for some the only), textbooks have to take most of the burden of conveying the principles and the philosophy of the new curriculum.

It is in this backdrop that a study was conducted to understand the impact of the new curriculum framework and the new textbooks on the teaching and learning of mathematics in classrooms. The study was restricted to the area of algebra for grades 6 and 7, where one finds a gradual introduction to the use of conventions and dealing with an abstract symbolism to communicate one's understanding. In this chapter, I will explore what sense students are making of introductory algebra as a part of the new textbooks. In the process, I will try to evaluate the extent to which the visions of NCF-2005 are being fulfilled in terms of the content being taught, teacher preparation and classroom transaction of the content.

Studies on Algebra Learning

As indicated in the previous section, mathematics education does not have a distinct disciplinary standing in this country. Also, there does not exist any systematic body of research in the area. However, post NCF-2005, there has been some writing on the philosophy behind the new curriculum framework, its impact on how knowledge would be constructed in classrooms in different subject areas and issues which are likely to arise as a result of its implementation (e.g. Paliwal & Subramaniam, 2006; Saxena, 2006; Batra, 2006). None of these address the area of mathematics. The new mathematics textbooks were reviewed, though not systematically analyzed, by Tripathi (2006) where she notes the marked improvement in the books for primary and secondary/ senior secondary but not at the middle school level where a large amount of the content receives the same treatment as earlier.

One can also draw from studies in some other countries in the past that have shown the difficulties in the implementation of the reforms. Such difficulties arise due to lack of adequate content knowledge among teachers, inability to reach all sectors to train teachers, unavailability of textbooks and resource materials, teachers' differential perception of the curriculum with respect to the load, the content covered, different interpretations of the textbooks and the materials, unawareness of new techniques of assessment and the motivations for change. (Research Advisory Committee, 1988; Remillard, 1999; Swarts, 1998; Bulut, 2007). The use of curricular materials involves "interpreting the meanings and intents of these resources" and "the enacted curriculum is more than what is captured in the official document" (Remillard, 1999, p. 317). Several studies have also indicated the influence of teachers' consciously or unconsciously held beliefs, notions and preferences on the success of curricular reforms (Thompson, 1988; Handal & Herrington, 2003; Yates, 2006; Choksi, 2007). Teachers thus respond differently to curricular changes in varying situations – some resisting it, some changing their practices superficially to match the new expectations without changing their fundamental beliefs about the content and nature of mathematics and teaching and learning process and some changing their practices for the better.

Some scepticism for any reform comes mainly due to the lack of sufficient research base (empirical or theoretical) for many of the recommendations (Research Advisory Committee, 1988). In the context of this study as well, one is apprehensive of the new way of introducing algebra in grade 6, knowing that there is no consensus on the 'best approach to algebra' (see for different approaches, Mason et al., 1985). Each approach brings with it some solutions and some difficulties and they are all important to grasp the content and purpose of algebra. It is therefore important to know the difficulties in teaching and learning of algebra as well as in the use of different approaches to introduce algebra while writing the textbooks. In contrast to the earlier textbooks which introduced algebra as a set of rules and conventions for manipulating symbols, the new textbooks use a pattern generalizing approach.

The change is a result of the view that rule based syntactic transformations do not carry any meaning for students while pattern generalization provides students a space to construct meaning for the symbols in algebra. On the other hand, Banerjee (2008) demonstrates a way to create meaning for algebraic symbols in the context of syntactic transformations. It uses students' knowledge of arithmetic (numbers and a sense of operations) to build a foundation for beginning symbolic algebra. The teaching approach was evolved over a series of trials and highlighted the structure of arithmetic expressions. Through the use of concepts like "term", "equality" and formulation of rules of transforming expressions using terms, it helped students studying in grade 6 to connect arithmetic and algebra. Further, students' understanding of transformation of expressions was used in contexts where algebra acted as a tool for generalizing, proving/ justifying. Analysis of students' written and interview responses as the approach evolved revealed the potential of the approach in creating meaning for symbolic transformations in the context of both arithmetic and algebra as well as making connections between arithmetic and symbolic algebra. Students by the end of the trials learnt to use their understanding of both procedures and a sense of structure of expressions to evaluate/simplify expressions and reason about equality/equivalence of expressions both in the arithmetic and the algebraic contexts.

Implications and Suggestions for Future Research

The nature of curriculum reform is multifaceted. A framework document and rewriting of the textbooks is not sufficient to change the teaching-learning process or radically improve students' understanding. Meaning does not reside in the activities themselves; one has to construct meaning out of activities and teachers play an important role in this process. Thus, there is a need for professional development of teachers, engaging them in the reform process – allowing them to build an understanding of the motivations behind the reform, the philosophy guiding the reform and the nature of change in the content, its implications for classroom environment. At the same time, the textbooks must be able to convey some of the principles of teaching the content. More importantly, research evidence must be collected for the ideas being used in the textbook and analyzed against other alternatives before they form part of the textbooks. The study must be carried out on a larger scale with a broader focus involving other content areas to assess the impact of the curricular reforms.

Keywords: algebraic thinking, learning of algebra in India

REFERENCES

Banerjee, R. (2008). *Developing a learning sequence for transiting from arithmetic to elementary algebra*. Unpublished doctoral dissertation, Homi Bhabha Centre for Science Education, T. I. F. R., Mumbai.
Batra, P. (2006). Building on the National Curriculum Framework to Enable the Agency of Teachers. *Contemporary education dialogue*, 4(1), 88-118.
Booth, L. R. (1984). *Algebra: Children's Strategies and Errors*. Windsor, UK: NFER-Nelson.
Bulut, M. (2007). Curriculum reform in Turkey: a case of primary school mathematics curriculum. *Eurasia Journal of Mathematics, Science and technology education, 3(3)*, 203-212.
Choksi, B. (2007). *Evaluating the Homi Bhabha Curriculum for Primary Science: In situ*. Technical report no. I (07-08), Homi Bhabha Centre for Science Education, Mumbai.
Handal, B., & Herrington, A. (2003). Mathematics teachers' beliefs and curriculum reform. *Mathematics Education Research Journal, 15(1)*, 59-69.
Hart, K. M., Brown, M. L., Kuchemann, D. E., Kerslake, D., Ruddock, G., & McCartney, M. (Eds.). (1981). *Children's Understanding of Mathematics: 11-16*. London: John Murray.
Mason, J., Graham, A., Pimm, D., & Gowar, N. (1985). *Routes to/Roots of Algebra*. Walton Hall, Milton Keynes: The Open University Press.

National Council for Educational Research and Training (2005). *The national focus group on the teaching of mathematics: Focus Group Paper.* New Delhi: NCERT.

Paliwal, R., & Subramaniam, C. N. (2006). Contextualising the Curriculum of the Poor: Concerns Raised by the 'Blocked Chimney Theory. *Contemporary education dialogue,* 4(1), 10-24.

Remillard, J. T. (1999). Curriculum materials in mathematics education reform: a framework for examining teachers' curriculum development. *Curriculum Inquiry, 29(3),* 315-342.

Research Advisory Committee of the National Council of Teachers of Mathematics. (1988). NCTM curriculum and evaluation standards for school mathematics: responses from the research community. *Journal for Research in Mathematics Education, 19(4),* 338-344.

Saxena, S. (2006). Questions of Epistemology: Re-evaluating Constructivism and the NCF 2005. *Contemporary education dialogue,* 4(1), 25-51.

Swarts, P. (1998). Evaluation and monitoring exercise of the mathematics curriculum. Executive summary of the report. National Institute for Educational Development, Namibia.

Tompson, A. G. (1988). The relationship of teachers' conceptions of mathematics and mathematics teaching to instructional practice. *Educational Studies in Mathematics,* 15(2), 105-127.

Yates, S. M. (2006). *Primary teachers' mathematics beliefs, teaching practices and curriculum reform experiences.* Paper presented at the AARE conference, Adelaide: Australia.

CHAPTER 70

PROFESSIONAL DEVELOPMENT OF IN-SERVICE MATHEMATICS TEACHERS IN INDIA

Ruchi S. Kumar, K. Subramaniam, and Shweta Naik
Homi Bhabha Centre for Science Education (TIFR), Mumbai

EXTENDED ABSTRACT

Teachers are central to any education system. There is growing realization across the world that any reform in education cannot be brought about without adequately addressing teachers' role in it. Teachers in India at present face huge challenges as India struggles to implement the National curriculum Framework (NCF) 2005 and the Right to Education Act (RTE) 2009 in schools. These challenges include providing them quality education through student centered pedagogy, trying to understand students' milieu, integrating ICT with teaching, assessing students comprehensively and continuously, and communicating with parents among other important tasks. The vision of teaching given by NCF 2005 is progressive but underestimates the system wide preparation that is needed to bring about the intended reform. Changing textbooks and passing circulars is not enough to sensitize teachers, administrators and parents towards the vision of active learning and independent learner as portrayed in the new curriculum framework. NCF 2005 is silent on how teachers are supposed to bring about the change in their classroom and does not address the much needed teacher development to support curriculum renewal (Batra, 2005).

Studies have indicated that the present teacher education, pre-service as well as in-service, fails to address the needs of the teachers as they do not consider the stark realities that teachers face in their schools like large class size, multilingual students and incomplete understanding of the content in the textbook (Ramachandaran et al., 2009). The problems in the design and mode of teacher education and professional development programs has been recognized and change has been recommended by several national committees, but little change has come over the years (Batra, 2005). The new National Curriculum Framework for Teacher Education (NCFTE, 2009) defines the relationship between teacher education and qualitative improvement in school education as a "symbiotic" relationship and recommends "urgent and comprehensive" reform in teacher education.

In order to design better in-service Teacher professional development (TPD) programs, there is need to develop clear vision of the goals that TPD programs must achieve and the means by which they can be achieved. Most in-service TPD programs in India are designed in response to the need of curriculum reform and view teachers as 'agents' of the state. Underlying this is the assumption that teaching can be changed by altering the content/structure of interactions in classrooms while not directly addressing teacher's own conceptions of teaching, learning and mathematics. In-service TPD is seen as training for content or pedagogy, mostly revolving around the changed curriculum, but not necessarily as important for continuous teacher development.

There can be different types of professional development opportunities for teachers like action research, case inquiry, narrative inquiry, focus on student thinking, model lessons and workshops (Zaslavsky et al., 2003). In India, workshops are an important component of TPD programs on which maximum time, effort and resources of the state are spent. In our experience, TPD workshops are often organized in an ad hoc manner on the basis of expediency, sometimes driven by the need to utilize funds. There is no clear consensus about what needs to be done in these workshops and how it is to be done. Resource persons, who are chosen by the coordinators on the basis of availability and willingness, design and conduct their sessions without considering how their session fits into the workshop as a whole. In structured large-scale programs, TPD is sought to be achieved through the "cascade model" of training, where master resource teachers get trained first and they do the training of the teachers from the next tier. The design and content of the modules, which are used repeatedly at each tier of the cascade training, is generally not based on empirical evidence. The vision underlying most of these programs restrict teachers' agency to implementing a new textbook, a pre-designed pedagogy or a prescribed assessment technique.

One needs to understand the beliefs, conceptions and practices of teachers as well as understand the contexts in which they are placed, in order to design a program that would be effective and useful. Thus an important criteria that in-service TPD programs need to meet is to bring the workshop goals closer to the actual work of teaching and help in developing teachers as professionals, creating opportunities to put them in-charge of their own learning.

With the passing of the Right to Education Act, and the consequent pressure to universalize elementary education, most states are faced with a shortage of teachers. This has led to multiple cadres of teachers and the appointment of "para-teachers" with lower educational qualification than that required of a regular teacher. This policy measure reiterates the assumption that a primary teacher does not need to know mathematics beyond the level that he/she is going to teach. Thus there are very low expectations by policymakers regarding the level of content knowledge required of a primary teacher. Various studies (Banerji & Kingdon, 2010; Ravindra, 2002-03; Dewan, 2009) have revealed the unsatisfactory status of knowledge of mathematics of even regular school teachers raising skepticism about the knowledge of para-teachers. This state is a reflection of teachers' own education which valued only rote memorization of procedures on the one hand and lacked opportunities to relearn mathematics in a meaningful way during professional education and during the course of their career on the other hand.

With the change in curriculum, the demand for better understanding of the content and alternative pedagogy has increased. Teachers in elementary and middle grades not only have to make their students fluent in computational mathematics but also need to make them proficient in practices associated with doing mathematics, such as reasoning, using multiple ways to solve problems, justifying their solution, making generalizations and conjectures, analyzing mathematical work of others, etc. However there have been few TPD programs in India, which have focused on the skills and knowledge required to facilitate this kind of teaching. Pedagogical content knowledge (PCK) (Shulman, 1986) is a specialized knowledge required for teaching of mathematics and subject matter knowledge (SMK) is a coherent, connected and deep understanding of mathematics (Ma, 1999). Although SMK and PCK are widely acknowledged now as essential components of teachers' knowledge, the preparation of content and pedagogy revolving around content is rarely the central focus of any phase of teacher education. There is no reason to assume that our teachers enter the profession with good SMK and PCK, as there is no exposure to such knowledge in their learning trajectory. This lays a great demand on TPD programs to provide opportunities for teachers learning mathematics and pedagogy revolving mathematical practices.

Bringing about change in teachers' knowledge of mathematics relevant to teaching is clearly a challenging task. Studies have shown that teachers' beliefs also strongly influence teaching practice and determine what teachers notice in the classroom (Thompson, 1992). In India most teachers' teaching is shaped by what is expected of them in the system as well as the kind of folk ideas of teaching that teachers as a community have a shared understanding of (Ramachandaran et al., 2009). Prema Clarke's (2001) work is illustrative of the cultural models of pedagogy that teachers hold, the way they perceive their roles and responsibilities in the classroom and how they convey it to students through their words and behavior. A study by Dewan (2009) indicates that beliefs held by not only teachers but even administrators, faculty members and directors of teacher education institutions are far removed from the ones envisioned in NCF 2005, thereby indicating the extent of challenge to implement the new framework. It is necessary to create spaces where teachers as well as other stakeholders articulate and reflect on the beliefs that they hold. Thus professional development programs should include opportunities to share and discuss views held by teachers while respecting their identity as a professional. Studies identifying teachers' beliefs and how they are held will help to design TPD programs to allow teachers to articulate and reflect on their beliefs rather than just superficially adopting the pedagogies proposed.

In this paper we address the concern about the focus and design of TPD programs in the context of school mathematics teaching in India. We situate our discussion in the background of recent research on teacher education that has thrown light on how TPD can be related to the work of teaching. Many of these research studies have developed tasks that address a variety of concerns central to TPD programs. We illustrate how these tasks may be integrated into a framework through a discussion of the TPD programs that the authors have been involved in.

The TPD program at the Homi Bhabha Centre with primary and middle grade mathematics teachers, had a component of workshops as well as the aspect of follow up in classrooms through collaboration with teachers. The workshops addressed the need for development of knowledge and beliefs conducive to teaching for understanding. The aim was to aid reflection and provide opportunities to engage in tasks that helped them analyze teaching in terms of the various aspects contributing to decision making. In this chapter we present the framework, that guided our design of the workshop sessions and the selection of tasks for teachers. We believe that it is important to not only consider what the content of the program was but also 'how' it was enacted. We present an analysis of two workshop sessions chosen to illustrate the aspects of facilitation whereby the understanding of mathematics teaching was negotiated. The workshop sessions were video recorded and descriptions of the sessions were generated using these videos by the authors. These descriptions were then coded for the design features, the facilitation features and the teacher explorations and reflections. Reliability for the codes was achieved through discussions among the authors and the development of consensus. Our preliminary findings reveal features of a potentially fruitful TPD workshop design where activities that are content based and associated with teaching practice provide opportunities for teacher learning.

Keywords: teacher professional development, India, mathematics teachers

REFERENCES

Hill, H. C., Sleep, L., Lewis, J. M., & Ball, D. L. (2007). Assessing teachers' mathematical knowledge: What knowledge matters and what evidence counts? In F. K. Lester (Ed.), *Second handbook of research on mathematics teaching and learning* (pp. 111-155). Charlotte, NC: Information Age Publishing.

Batra, P. (2005).Voice and Agency of Teachers: The missing link in the National. Curriculum Framework 2005, *Economic and Political Weekly,* 40(36), 4347-4356.

Clarke,P. (2001). *Teaching and Learning-the culture of pedagogy.* New Delhi: Sage Publications.

Dewan, H. K. (2009) Teaching and Learning: The Practices. In Sharma, R. & Ramachandaran, V. (ed.). *The elementary education system in India. Exploring institutional structures, processes and dynamics.* New Delhi: Routledge.

Banerji, R., & Kingdon, G. (2010) "How Sound Are Our Mathematics Teachers? Insights from school Tells survey", *Learning Curve,* No. XIV (pp. 52-56).

Ma, L (1999). *Knowing and teaching elementary mathematics: Teachers' understanding of fundamental mathematics in china and the united states (Studies in mathematical thinking and learning,* Lawrence Erlbaum associates: Mahwah, NJ.

NCF (2005). National Curriculum Framework , NCERT, New Delhi.

National curriculum framework for teacher education (2009) *Towards preparing professional and humane teacher.* NCTE, New Delhi.

Ramachandran, V., Bhattacharjea, S., & Sheshagiri, K. M. (2009) *Primary School teachers. The twists and turns of everyday practice.* Educational resource unit, New Delhi.

Ravindra, G. (2002-03) Research on curriculum and teaching mathematics, In *Sixth survey of educational research 1993-2000,* Vol II, (pp 362-376), NCERT, New Delhi.

Shulman, L. S. (1986). Those who understand: Knowledge growth in teaching. *Educational Researcher* Feb. 1986: 4-14. (AERA Presidential Address).

Thompson, A. (1992). Teachers' beliefs and conceptions:A synthesis of the research. In D. A. Grouws (Ed.), *Handbook of research on mathematics teaching and learning* (pp. 127-146). New York: Macmillan.

Zaslavsky, O., Chapman, O., & Leikin, R. (2003) Professional development in mathematics education: Trends and tasks. In A. Bishop et al. (Eds.), *Second international handbook of mathematics education* (875-915). Dordrecht, The Netherlands: Kluwer.

CHAPTER 71

INSIGHTS INTO STUDENTS' ERRORS BASED ON DATA FROM LARGE-SCALE ASSESSMENTS

Aaloka Kanhere
Homi Bhabha Centre for Science Education (TIFR), Mumbai

Anupriya Gupta and Maulik Shah
Educational Initiatives Pvt. Ltd, India

EXTENDED ABSTRACT

Students bring ideas from real-life experiences or notions built in earlier grades to the classroom. Some of these ideas help the student to further build on her knowledge whereas some hamper her understanding. In this chapter we look at students' errors and notions which can form hurdles to developing a clear understanding of mathematics. These errors can be of various types such as errors which stem from wrong notions about mathematical concepts, or incorrect recollection of rules, or generalizing already known results to an inappropriate situation. In this chapter we look at a broad category of errors but exclude slips or careless errors which any student, or even experts, might commit without any rationale.

An in-depth understanding of errors and underlying student ideas, their prevalence across grades, and the likelihood of their occurrence in different groups of students is a critical part of a teacher's pedagogical content knowledge (Shulman, 1987).

In India, the research on misconceptions so far has been largely based on classroom observations and interviews with individual students. With the improvement in data collection technologies in the last two decades, it has been possible to collect and analyse the responses of student populations to the tune of tens of thousands for a large number of questions across grades. Benchmarking studies on student learning standards and diagnostic tests in each subject are also being carried out globally (Loveless, 2007). The data from all these studies serve as a repository for finding common patterns in errors that students make, and provide granular information of student knowledge. Data on student errors is also collected by Intelligent Tutoring Systems (ITS) and used also as formative assessment tools to give records of student learning processes (Pellegrino, Chudowsky, & Glaser, 2001). An advantage of data collected from an ITS is that it provides across-class level data for items and allows comparison of student errors at different age levels.

For the scope of this chapter, the data available is mostly in the form of responses on multiple-choice questions answered by students of varying grade levels (3 – 10). On some items, response data is available for about 25,000 students. For the purpose of this chapter, the student response data from the following sources have been used:

- ASSET—developed by Educational Initiatives Pvt. Ltd., is a diagnostic test taken by about 450,000 students every year across different classes of private English medium schools in India. All the students of different ability groups of participating schools take the test. The data is taken from the tests taken by students in the last 5 years.
- Mindspark (ITS)—developed by Educational Initiatives Pvt. Ltd., is a computer based adaptive self-learning programme taken by private English medium schools in India.
- Data from benchmarking studies such as Student Learning Study (SLS)[1] (2009), SLIMS[2] study (2006) and Wipro Quality Education Study[3] (2011).

The analysis of the data from these sources can confirm or contradict the findings from small-scale studies leading to more fine-grained insights about student errors and the pedagogy of the specific topics, especially in the Indian context.

In India, the research in mathematics education on student errors is mostly done in the context of interventions in improving classroom instruction. For instance, Menon lists the difficulties faced by students when they are first introduced to the topic of 'Angles' in school (Menon, 2009) as a part of her work on the development of an alternative trajectory to teach geometry. One of the student errors stated by her is that children tend to associate the measure of an angle with the length of its arms. The book, Compendium of Errors in Middle School Mathematics by Pradhan and Mavalankar provides a limited compilation of student errors (Pradhan & Mavalankar, 1994).

This literature can be enhanced if our understanding about students' errors can be improved by studying the prevalence of such errors over large student populations. It is also important to know if different student-populations show different error-patterns. The persistence of errors across learning levels is also a useful insight for math educators.

Examples of Errors and Underlying Student Thinking

In this section, we first present an example of a common student error in the topic of "Area". An Item Response Curve (IRC) generated on the data of a large scale diagnostic test (ASSET) shows the extent of the error in different ability groups of class 7. Further, we analyse the prevalence of incorrect notion in the topic "Angles" across classes. It was seen that though certain incorrect notions get cleared in higher classes, others remain strong, and continue to create learning difficulties for students.

An Incorrect Notion About "Area"

Students think that irregular shapes made up of curved lines don't have an area. Performance of students is presented in the Table 71.1. The correct answer is Option D and the most common wrong answer is Option C. As per the guidelines of National Curriculum Framework 2005 (NCF) the concept of area is introduced informally in Class 5 and more systematically in class 6. The data above suggest that even in class 7, about 37% of the students across ability-levels think that shapes with curved lines don't have area. It can be seen in the IRC graph (Figure 71.2), that about 20% of the students who answered three-fourth of all 40 questions correctly, have this idea. Thus this idea exists even among a few high-performing students.

Such information, about whether a particular incorrect notion exists only in certain groups of students, or across ability levels can help the teacher identify target groups for providing further instruction.

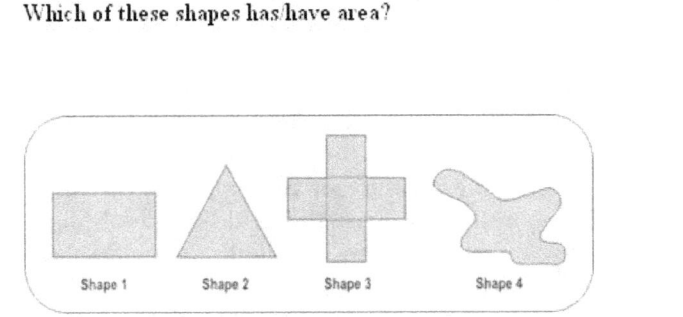

A. only shapes 1 and 2 because the area can be calculated using formulae
B. only shapes 1 and 3 as they are the only ones which can be covered completely by unit squares
C. only shapes 1, 2, and 3 as they have straight sides
D. all four shapes - 1, 2, 3 and 4

Figure 71.1.   Class 7, Question 1.

Table 71.1.

	No. of Students	Option A	Option B	Option C	Option D
% students	25371	13.6%	13.0%	37.1%	34.8%
PBC[a]		-0.22	-0.14	-0.08	0.37

[a]Point-biserial correlation coefficient (PBC) is a correlation coefficient used when one variable is dichotomous. Larger the value of PBC of answer option, the more is the discriminating value of the item. Here the PBC values are inclusive of the given item. Negative PBC for the correct answer option indicates that item may not be functioning.

Prevalence of Students' Errors Across Classes

Research indicates that some students, with an incorrect notion about the 'measure' of an angle, compare the lengths of the arms when asked to compare angles (Menon, 2009). However, the quantitative data supporting this claim was not given in this study. The data from Mindspark confirms the existence of this notion and adds to our knowledge about its nature.

Abstracts of The First Sourcebook on Asian Research in Mathematics Education 223

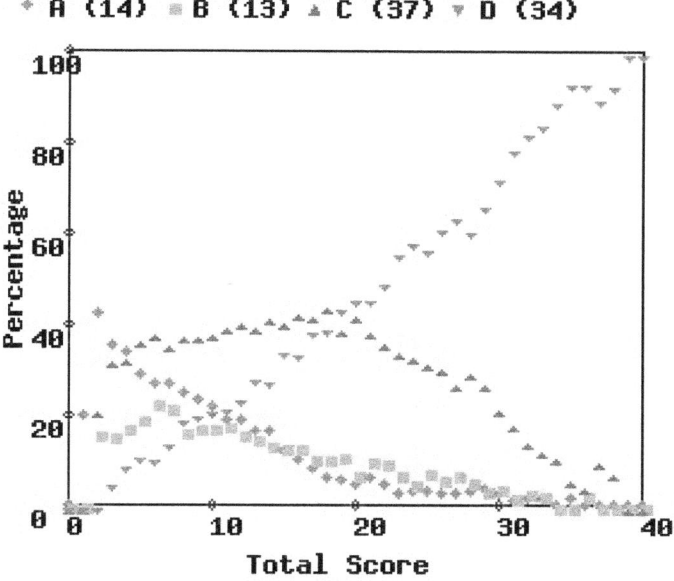

Figure 71.2. This data for the question in Figure 71.1 is of Class 7 students from various English-medium private schools across India.

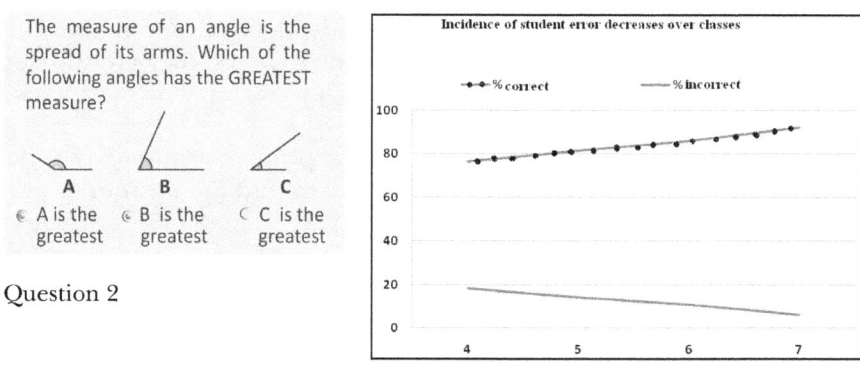

Question 2

Figure 71.3
This data is of students from various English-medium private schools across India.

When asked Question 2, 18% of students of Class 4 chose Option B, the angle with the longest arms. In Class 7 only 6% of students chose that

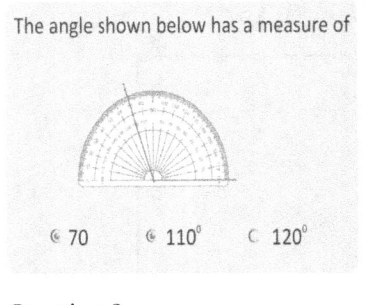

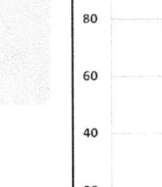

Question 3

Figure 71.4.

This data is of students from various English-medium private schools across India.

option. This suggests that the understanding of the measure of an angle improves over class levels as desired.

The skill of visually estimating the measure of an angle helps students in measuring an angle using the protractor. When checked with other examples it is evident that even though students learn to compare the sizes of angles fairly well as they move to higher classes, they are still not able to measure the angle correctly using the protractor. The example in Figure 71.4 illustrates this point.

In this question, a student has to move the protractor and place it on an angle and read the scale on the protractor to find the measure of the angle shown. Around 20% of the students of class 7 are found to struggle while reading the scale on the protractor. The persistence of this error as well as a high proportion of correct responses from the early grades suggests possible strategies and times of intervention.

The knowledge about which incorrect notions in a topic are likely to persist across classes, and which are not, can be powerful information for the teacher – who often needs to make decisions on which area to focus on (and which to defocus) during classroom interactions (Srinivas, 2010).

Future Research Implications

We plan to further analyse the evidence of the certain student errors mentioned in this article and explore the roots of incorrect notions students have. It is foreseen that such analysis could bring about changes in pedagogical approaches to teaching, in the curriculum, and it can help in lowering the occurrence of incorrect notions among students.

Keywords: student error analysis, India, mathematics education research, assessment, evaluation

NOTES

1. The project undertaken by Educational Initiatives supported by Google involves assessment of student achievement in classes 4, 6 and 8 and covers about 21 states and thus about 15 languages. It aims to establish student achievement levels and to provide insights into comparative performances of different states, to create cross-learning and remedial opportunities.
2. Educational Initiatives (EI) and Wipro Ltd. conducted this research study in the 5 metro cities – Mumbai, Kolkata, Chennai, Delhi and Bangalore among the 30,000 students from 'top' 40 schools in 2006. (http://www.ei-india.com/whats-wrong-with-our-teaching)
3. This is an ongoing research study undertaken by Educational Initiatives (EI) and Wipro Ltd. in the 5 metro cities – Mumbai, Kolkata, Chennai, Delhi and Bangalore and a few schools with different learning environment.

REFERENCES

Loveless, T. (2007). Chapter 6 -What can TIMSS Surveys Tell Us About Mathematics Reforms In the United States during the 1990s. In *Lessons Learned: What International Assessments Tell Us About Math Achievement.* Washington, DC: Brookings Institution Press.

Menon, U. (2009). The Introduction of Angles. *Proceedings of epiSTEME-3 International Conference to Review Research on Science, Technology and Mathematics Education* (pp. 133-138). Mumbai: Macmillan.

Pellegrino, J. W., Chudowsky, N., & Glaser, R. (2001). *Knowing what students know: The Science and Design of Educational Assessment.* Washington: National Academy Press.

Pradhan, H. C., & Mavalankar, A.T. (1994). *Compendium of Errors in Middle School Mathematics,* Mumbai: Homi Bhabha Centre for Science Education.

Shulman, L. S. (1987). Knowledge and teaching: Foundations of the new reform. *Harvard Educational Review* (pp. 1-22).

Srinivas, S. B. (2010). Mining information from tutor data to improve pedagogical content knowledge. *The Third International Conference on Educational Data Mining (EDM2010)* (pp. 275-276). Pittsburgh.

CHAPTER 72

ASSESSMENT OF MATHEMATICAL LEARNING

Issues and Challenges

Shailesh Shirali
Rishi Valley Education Centre, Andhra Pradesh

EXTENDED ABSTRACT

Challenges Posed by Educational Assessment in India

It is axiomatic that assessment should be integrally linked with learning, that it not be merely a device to measure cumulative learning at the end of a teaching unit. As one farmer pithily remarked at a public hearing in USA, "You can't fatten a hog by weighing it" (Measuring What Counts, 1993). In our context, we may say that summative assessment does not add educational value, and that its sole benefit (if any) lies in its lateral action. But the sad reality in a good many parts of the world is that "assessment" is precisely that; it is limited to the summative kind, and is purely a device to measure cumulative learning, used at the tail end of each academic year, a device to help the teacher write reports, a device to help make pass/fail decisions.

This is particularly the case in India. In very few countries is it as perniciously true as here that summative assessment holds the key to one's future, in the sense of opening or closing doors of opportunity; in the latter case, closing them very tightly indeed. The problem is of sufficient gravity that every year there are suicides associated with it: children unable to cope with the disappointment and shame of failure, or with the fear of condemnation. Inevitably the specter of such assessment exercises a significant influence on the ambient educational culture, inviting poor educational practices and the creation of a powerful parallel education system called 'coaching' (or 'shadow education system'). Indeed, it invites criminal activity as well, through the leakage and subsequent sale of examination papers. The last implication is a significant area of concern in Indian education today.

Looking closely, we see four powerful inter-related forces that have a bearing on assessment:

(a) The culture and mindset of the teacher community;
(b) The style and approach followed by the textbooks;
(c) The culture of the examination system;
(d) The curriculum itself.

In this study we only consider (c) and, by implication, (d); the other two – (a) and (b) – are major fields of study in themselves. Thus our focus is on the question: *How is assessment in mathematics currently handled by schools and examination boards across the country?*

Included in this are queries such as the following: *What is the relation between assessment and the level of conceptual understanding and mastery of skills? What place do higher order thinking skills ('HOTS') have in mainstream assessment?* We touch on these briefly.

We begin the section by describing the role played by the National Council for Educational Research and Training (NCERT) in the educational scenario of India, mentioning the document it released in 2005, the National Curriculum Framework (NCF); follow this with an overview of the educational system in the country: the many State examination boards and the national boards, and the challenges posed by the existence of so many independent bodies. We then dwell on a problem peculiar to India – the problems posed by college admission and the phenomenon of the tutorial colleges ('coaching centres'), whose methodologies have had a significant negative influence on the ambient educational culture of the country.

Assessment Studies Carried Out in India—
A Survey of the Existing Literature

Some of the documents that we study in this section are listed in the references. We describe the NCF 2005 and the comments as well as recommendations it makes about assessment. Then we report on two studies: one made by PISA (Turner, 2010) and the other by Educational Initiatives (Student Learning in the Metros, 2006); the latter brings out misconceptions that persist in students' minds even in some so-called "good" schools.

Study of the Approaches Followed
by Examination Boards in India

In the next section we study the way a well known national examination board (the Council for Indian Schools Certificate Examination or CISCE) and two very well known competitive examination agencies (the Joint Entrance Examination of the Indian Institutes of Technology and the All India Engineering Entrance Examination, which carry extremely high stakes) deal with some typical topics and with problem solving in general. We examine questions from a range of topics so as to obtain a picture of the mathematics covered by these examinations. Here are the topics we look at.

- Public examinations, high school level (Class 10): Computation of Value Added Tax (and commercial arithmetic in general), Trigonometry, Algebra (polynomials, equations), Probability, Descriptive statistics: mean, mode, median
- Public examinations, senior secondary level (Class 12): Matrix algebra, Coordinate geometry, Differential and integral calculus, Boolean algebra, Descriptive statistics: correlation, regression and moving averages, Mean value theorem and L'Hospital's rule for limits of indeterminate forms

1. Competitive examinations, IIT-JEE and AIEEE: Differential and integral calculus, Coordinate geometry, Limits, Combinatorics

Typically we find that students are expected to possess manipulative abilities of a high order. Example (asked in a Class 12 exam):
Find the derivative of the following with respect to x:

$$\tan^{-1}\sqrt{\frac{a-x}{a+x}}.$$

Occasionally we also find a few trick questions which can likely be solved only if they have been seen earlier. Here is an example of such a question (again, from a Class 12 exam):

Evaluate the integral $\int \frac{xe^x}{(x+1)^2}dx$.

We also comment on the fact that some topics such as Descriptive Statistics draw forth a very mechanical type of question, simply asking for the computation of a few test statistics with no interpretation of any kind attached to them.

A finding that emerges is that pattern recognition and speed play a key role in the competitive examinations. This is the area where the coaching centres score heavily over "raw" untutored ability.

Implications and Suggestions for Future Research on Assessment

The NCF document states: "… while there has been a great deal of research in mathematics education and some of it has led to changes in pedagogy and curriculum …, the area that has seen little change … over a hundred years is evaluation procedures in mathematics. It is not accidental that even a quarterly examination in Class VII is not very different in style from a Board examination in Class X, and the same pattern dominates even the end-of chapter exercises given in textbooks. It is always application of some piece of information given in the text to solve a specific problem that tests use of formalism. Such antiquated and crude methods of assessment have to be thoroughly overhauled if any basic change is to be brought about …."

With this statement as backdrop, here are some directions we envisage (and recommend) for future educational research in the country:

1. *Empirical studies on diagnostic and normative assessment.* In the mainstream of the country, there is low awareness of the potential of diagnostic assessment. We need to take steps which will demonstrate its potential and thus bring it into the mainstream. There is similarly great resistance to normative testing; currently, TIMMS and PISA are almost unknown in the country. We need to embrace

such international practices. A study of their pedagogical utility would have great value.
2. *Scope and effects of computerized testing, and the possibility it holds for diagnosing and identifying gaps in understanding.* Included here would be the matter of gender bias involved in multiple choice testing. Also, whether students with difficulties in the subject benefit from low stakes computerized testing because of the reduced level of anxiety and the possibility of repetitive testing.
3. *Scope of investigative projects, and more generally of open-ended work.* What would be a fair way of assessing such work? How can concerns for integrity be addressed? (Integrity of assessment is currently a problem of major dimensions in India. The high stakes nature of public and competitive examinations inevitably brings them under the gaze of criminal activity.)
4. *Effects on conceptual understanding, originality and creativity in mathematics as a result of high intensity examination coaching during the formative school years.* (This includes coaching for the competitive examinations which we mentioned above.)
5. Incidence of dyscalculia in urban and rural India and how high stakes examinations affect those with such difficulties.
6. Effects on mathematical abilities brought about by the use of calculators in school, including graphics calculators.
7. The interplay between teacher training, teacher attitudes, and patterns of assessment.

Keywords: competitive examinations, university entrance, assessment, evaluation, India mathematics exams

REFERENCES

CBSE HOTS questions. (2011). Retrieved from http://www.icbse.com/2009/cbse-hots-question-paper-class-10-12/

Continuous and Comprehensive Evaluation. (2011). Retrieved from http://www.cbse.nic.in/cce/index.html

Education in India. (11 March 2011). Retrieved from http://en.wikipedia.org/wiki/Education_in_India

Gilderdale, C. (2007). Lower and Higher Order Thinking. Retrieved from http://nrich.maths.org/5795

KVPY Eligibility SP (Basic Sciences). (2011). Retrieved from http://www.kvpy.org.in/main/eligibility-spbasicsciences.htm

Measuring What Counts. (1993). Mathematical Sciences Education Board, National Research Council, USA (1993). Retrieved from http://www.nap.edu/catalog/2235.html

National Curriculum Framework. (2005). Retrieved from http://www.ncert.nic.in/rightside/links/pdf/framework/nf2005.pdf
Student Learning in the Metros. (2006). How well are our students learning? Retrieved from http://www.ei-ndia.com/full-report.pdf
Turner, R. (2010). Lessons From The International PISA Project. *Learning Curve.* (Azim Premji Foundation). March 2010, 84–87.

CHAPTER 73

TECHNOLOGY AND MATHEMATICS EDUCATION

Issues and Challenges

Jonaki B. Ghosh
Lady Shri Ram College for Women, University of Delhi

EXTENDED ABSTRACT

Mathematics has for years been the common language for classification, representation and analysis. Learning mathematics forms an integral part of a child's education. Yet, it is also the subject, which has traditionally been perceived as difficult. The primary reason for this state of mathematics learning today is the significant gap between content and pedagogy. The last few decades have witnessed serious experimentation and research in mathematics education all over the world and there has been a shift of paradigm as far as mathematics teaching and learning is concerned. Mathematics education is being revolutionized with the advent of new and powerful technological tools. Because of these tools mathematics education can focus on problem solving, reasoning and communicating that empower students to confidently explore, conjecture and reason logically. While traditional mathematics is often fraught with rote memorization of procedures, computational algorithms, paper-pencil-drills and manipulation of symbols, the use of technology has the potential to encourage teachers and students to engage in deep mathematical thinking involving analysis, problem posing, problem solving and rich conceptual understanding. Countries all over the world are increasingly using technology in mathematics teaching and learning but the same in not the case in India. Although integration of technology in schools is not uncommon in India, its use in mathematics teaching and learning is not prevalent. This chapter highlights the many challenges facing school mathematics education in India and suggests ways in which technology can play a role to overcome these challenges. The chapter also focuses on the challenges of integrating technology in the mathematics curriculum. The use of mathematics laboratories as a way of integrating technology has been suggested based on a project conducted by the author.

Challenges of School Mathematics Education in India.

The challenges facing school mathematics education in India may be broadly categorised under the following heads.

Transaction of Curriculum

Mathematics in school is taught as an abstract subject and for the vast majority of Indian students mathematics as taught at the school has no relevance to real life. Topics are taught without any substantial reference to their applications to real life problems which renders mathematics a dry and uninteresting subject. Traditionally the emphasis has been on routine problem solving and on the development of 'by hand' skills.

The curriculum is disappointing for the talented minority as well as the average or below average majority. Over the years curriculum coverage has been enhanced by adding new topics or removing certain topics but there has been no major change in the approach to dealing with these topics. Mathematics suffers from being the most feared subject and often is the reason why many students drop out of school.

The curricula followed by the majority of schools do not permit the use of technology for the teaching and learning of mathematics. The chalk and board is predominantly used in mathematics classrooms and the use of scientific or graphic calculators and software packages is not encouraged at any level.

Inappropriate Assessment

The school mathematics curriculum suffers from the ailment of obtaining 'the one right answer'. "Exactness" is overemphasized and problems posed in examinations test the ability to obtain the "right" answer. Thus the focus is on the development of procedural knowledge and memorization of formulae and examinations also test the same. Examinations conducted by the central and state boards, which are high stakes examinations, end with the awarding of marks or grades to the student by which she tends to get branded for life.

Teacher preparation

Teachers are indeed the most critical resource in the system. In India however a vast section of mathematics teachers lack the opportunity to equip themselves with modern means to teach the subject imaginatively due to lack of proper training and development. Teachers who have taught for several years find little opportunity for professional development and up-gradation. There is a strong need for sustainable, structured professional development opportunities for teachers.

Goals of Mathematics Education

The National Curriculum Framework (NCF 2005) envisages two kinds of goals for mathematics education in India. The narrow goal is that of turning out employable adults who can contribute to society. This implies that the curriculum focus on numeracy. The broader or higher goal is focused on developing the inner resources of the student, that is, logical thinking, handling abstractions, etc. The latter indicates the need for a curriculum which

(a) Enables students to learn mathematics for logical thinking, representation, analysis and communication.
(b) Emphasizes the relevance of mathematics in solving real life problems.
(c) Motivates teachers to facilitate student's learning using a variety of methods and pedagogical techniques.

This implies that school mathematics becomes accessible to students of all levels of ability and also becomes more 'activity oriented'. There are three major challenges in attaining these goals.

Accessibility: Mathematics education must be cost effective and affordable by all. It should be easy to deploy to the masses. There are about 1.2 million recognized schools, 5 million teachers of which around 1 million are mathematics teachers.

Diverse curricula: Each state has a different curriculum. There are 31 state boards and three national boards namely, Central Board of Secondary Education(CBSE), Council for Indian School Certificate Examinations (ICSE), National Institute of Open Schooling (NIOS). The challenge is to bring all these boards to a common level. The problem is magnified by the fact that each state has its own language.

Teacher Preparation: The country is very short of good mathematics teachers. There is a pressing need for pre-service training programmes for mathematics education. In service teachers also need to undergo a change of mindset. Their motivation levels have to be raised through sustainable professional development. Professional development programmes should provide opportunities for peer learning where teachers can engage in discussions and share resources. There is a need for agencies which create resources and impart training through workshops focusing on raising the pedagogical content knowledge of teachers.

Role of Technology in Mathematics Learning

This section will focus on some of the theoretical frameworks that support the use of technology in mathematics teaching and learning. In many countries across the world, mathematics education has been revolutionized due to the influence of technology. Technology primarily includes computer software such as computer algebra systems (CAS), dynamic geometry software (DGS) and scientific as well as graphic calculators. The primary role of technology in teaching and learning mathematics may be broadly classified as follows.

Visualization of Concepts. Visualization is a key factor in mathematics learning and can be graphic, numeric or algebraic. Technological tools such as computer algebra systems (Herwaarden & Gielden, 2002; Lagrange, 1999; Lindsay, 1995), dynamic geometry software and graphics calculators (Waits & Demanna, 1999; Kutzler, 1999) create new teaching possibilities and lead to visualization of concepts in a manner that would have been impossible using the traditional methods of teaching.

Exploration and Discovery. Technology can greatly increase the range of the students' mathematical activities by enabling them to explore concepts and discover results for themselves. Students also explore problems by writing their own programs, which enable them to think actively about the 'processes' they are implementing in solving the problems. This increases their ability to think analytically. They can spend more time in developing concepts and ideas thus developing mathematical maturity rather than spending time in the routine calculations (Arnold, 2004).

Emphasis on experimental aspects of mathematics. Through the use of technology activities may be designed so as to provide students with the opportunity to acquire skill in observing, exploring, forming insights and intuitions, making predictions, testing hypothesis, etc. However care must be taken that these skills must be acquired without neglecting the traditional aspects of mathematics such as proving, generalising, and abstracting.

Focus on applications of mathematics. Using technology, projects can be designed which focus on applications of topics in the curriculum thus allowing the subject matter to be taught in a more relevant manner.

Redefining the role of the teacher. In a technology-enabled environment, the role of the teacher has been found to be significantly different from that of the traditional classroom. Here the teacher spends less time in lecturing, setting examination papers and evaluating answer booklets. The teacher acts as a facilitator, "leading" the students to making their "discoveries" rather than providing the solutions on the blackboard.

The role of technology in mathematics education needs to be examined under the following three domains:

(i) Doing Mathematics

This domain deals with the process of mathematical activity and how it is being revolutionised by technology. Due to technology the focus of mathematics education will shift from the development of "by-hand" skills to the development of conceptual understanding and applications of mathematics. Technology will take over the routine calculations and the student can concentrate on a deeper exploration of the subject at hand (Arnold, 2004; Thomas & Hong, 2004).

(ii) Understanding of the Teaching and of Mathematics

This domain is concerned with the processes of learning and teaching concepts, skills and strategies in mathematics and its applications. Technology has profound implications here particularly as a potential aid to teaching and learning. Mathematics educators must consider in depth the teaching possibilities created by the computers and calculators.

(iii) Mathematics Curricula and Teacher Training

The integration of technology in the mathematics curriculum will require that learning objectives be redefined and new curricula be developed to meet these objectives (Aydin, 2005). The 'concepts before skills' approach will take precedence over the traditional 'by hand' skills approach (Heid, 1988; Stacey, 2001). However for the implementation of computers and graphic calculators into the curriculum, it is essential that teachers be trained in the use of technology. This would mean extensive professional development programmes for teachers including workshops and training programmes.

Challenges in Implementing Technology in the Indian Context

Implementation of technology poses many challenges, the greatest being the socio-economic challenge. Priority of Government is to reach education to the masses. Technology must be cost effective and easy to deploy. The last few years have witnessed extensive use of computer technology in schools. However mathematics teaching continues in the traditional 'chalk and board' manner. Technology, if used for teaching mathematics, is primarily for demonstration purposes and does not involve the student actively. It is imperative that a mathematics curriculum be designed which integrates technology.

Pre-service teacher education programmes must be designed where student teachers are taught mathematics using various technological tools. Inservice teacher training programmes must focus on changing their mindset towards technology and on helping them overcome their 'technological anxiety'. Technology must play a role in developing their pedagogical content knowledge (Turnuklu & Yesildere, 2007).

The Laboratory Approach of Teaching Mathematics

This section will focus on the concept of mathematics laboratories and on how technology can be integrated into the curriculum through such laboratories. The laboratory approach of teaching mathematics provides students with the opportunity to 'discover' mathematics through exploration and visualization. In a laboratory students will be required to participate in activities which enhance their understanding of the subject as taught in the classroom and also provide a glimpse of what is beyond. The activities may comprise of various projects, experiments and modelling exercises based on the mathematics taught in the curriculum. The primary objective of any laboratory is to perform experiments and the same is true of a Mathematics Laboratory. An 'experiment' in mathematics may be described as an 'exercise' or 'project' which

1. Highlights some known concept based on a well known mathematical theory.
2. Sheds new light on some aspect of the topic being studied.
3. Leads to some original discovery on the part of the student.
4. Focuses on some interesting application of mathematics to a real life problem.

Technological tools such as computer algebra systems, dynamic geometry software, spreadsheets and graphics calculators can play a critical role in the activities of a mathematics laboratory. The remaining part of this section will focus on specific activities which were conducted by the author with students in a laboratory setting where various technological tools enabled students to explore, visualize and compute. These activities may be classified as follows

1. Visualization and exploration of concepts using various technological tools.
2. Simulation of problems in probability using spreadsheets.
3. Investigatory projects based on mathematical modelling and applications of topics taught in the curriculum.

The remaining part of the chapter highlights excerpts of laboratory modules conducted by the author in the above three categories with students from grades 9 to 12. In most of the modules, graphics calculators, spreadsheets or computer algebra systems form the vehicles of exploration. In all modules student's explorations and investigations are guided by carefully designed worksheets which enable the student to explore the concepts in a step by step manner. Students had to provide manual solutions in some parts of the worksheets which helped to maintain a balance between "by hand" skills and use of technology. Feedback taken from the students at the end of the modules revealed that they preferred the laboratory modules to their traditional classes and that their levels of interest in the topic could be sustained over longer periods of time because of technology. The author emphasises that in some modules technology served the purpose of a "mathematical investigation assistant" by taking over tedious computations leaving the students to focus on insights and concepts thus giving them more control over what they were learning. In other modules technology served the purpose of an "amplifier" by giving students access to higher level concepts which would not have been possible in a traditional teaching environment without technology. The role of the teacher was more of a facilitator guiding and scaffolding student's explorations. Overall the integration of technology enriched mathematics teaching and learning and opened up possibilities for both students and teacher alike.

Keywords: challenges of school mathematics education, mathematics laboratories, technology in India, teaching and learning mathematics with technology

REFERENCES

Arnold, S. (2004). Classroom computer algebra: some issues and approaches. *Australian Mathematics Teacher*, 60(2), 17-21.

Aydin, E. (2005). The use of computers in mathematics education: a paradigm shift from "computer assisted instruction" towards "student Programming". *The Turkish Online Journal of Educational Technology*, 4(2), 27-34.

Ghosh, J. (2003). Visualizing solutions of systems of equations using Mathematica. *Australian Senior Mathematics Journal*, 17(2), 13-28.

Ghosh, J. (2004). Exploring concepts in probability using graphics calculators. *Australian Mathematics Teacher*, 60(3), 25-31.

Ghosh, J. (2005). Visualizing and exploring concepts in calculus using hand held technology. *Proceedings of the 10th Asian Technology Conference in Mathematics (ATCM)*. Retrieved from http://www.atcminc.com/mPublications/EP/EPATCM2005/enter.shtml

Heid, M. K. (1988). Resequencing skills and concepts in applied calculus using the computer as a tool. *Journal for Research in Mathematics Education*, 19(1), 3-25.

Heid, M. K. (2001). Theories that inform the use of CAS in the teaching and learning of mathematics. *Plenary paper presented at the Computer Algebra in Mathematics Education (CAME) 2001 symposium*. Retrieved from http://www.lkl.ac.uk/research/came/events/freudenthal/3-Presentation-Heid.pdf

Herwaarden, O. V.,& Gielden, J. (2002). Linking computer algebra systems and paper-and-pencil techniques to support the teaching of mathematics. *International Journal of Computer Algebra in Mathematics Education*, 9(2), 139-154.

Kutzler, B. (1999). The algebraic calculator as a pedagogical tool for teaching mathematics. *International Journal of Computer Algebra in Mathematics Education*, 7(1), 5-23.

Lagrange J B (1999). *A didactic approach of the use of computer algebra systems to learn mathematics*. Paper presented at the Computer Algebra in Mathematics Education workshop, Weizmann Institute, Israel. Retrieved from http://www.lkl.ac.uk/research/came/events/Weizmann/CAME-Forum1.pdf

Lindsay, M. (1995, December). *Computer algebra systems: sophisticated 'number crunchers' or an educational tool for learning to think mathematically?* Paper presented at the annual conference of the Australasian Society for Computers in Learning in Tertiary Education (ASCILITE), Melbourne, Australia.

Stacey, K. (2001). *Teaching with CAS in a time of transition*. Plenary paper presented at the Computer Algebra in Mathematics Education (CAME) 2001 symposium. Retrieved from http://www.lkl.ac.uk/research/came/events/freudenthal/2-Presentation-Stacey.pdf

Thomas, M. O. J.,& Hong, Y. Y. (2004). Integrating CAS calculators into mathematics learning: partnership issues. *Psychology of Mathematics Education*, 4, 297-304.

Turnuklu, E. B.,& Yesildere, S. (2007). The pedagogical content knowledge in mathematics: Preservice primary mathematics teachers' perspectives in Turkey. *IUMPST, The Journal*. Retrieved from http://www.k-12prep.math.ttu.edu

Waits, B. K.,& Demana, F. (1992). The Power of Visualization in Calculus. In E. D. Laughbaum(Ed.), *HandHeld Technology in Mathematics and Science Education: A Collection of Papers* (pp. 49–67). *Columbus, OH: Teachers Teaching with Technology College Short Course Program at The Ohio State University.*

Waits, B. K.,& Demana, F. (1999). A New Breed of Calculator: They will change the way and what you teach!. In E. D. Laughbaum(Ed.), *HandHeld Technology in Mathematics and Science Education: A Collection of Papers*.

National Curriculum Framework. (2005). *Position paper of National Focus Group on Teaching of Mathematics*. New Delhi: National Council of Educational Research and Training.

CHAPTER 74

MATHEMATICS EDUCATION IN PRECOLONIAL AND COLONIAL SOUTH INDIA

Senthil Babu D.
Department of Indology, French Institute of Pondicherry

EXTENDED ABSTRACT

Mathematics Education in Precolonial South India

The aim of the chapter is to introduce the nature and characteristics of elementary mathematics education in south India during the eighteenth and the nineteenth centuries with particular reference to the Tamil region. This is part of an attempt to probe the possibilities of different regional mathematical traditions, which have their implications for contemporary mathematics education, not to mention its relevance to the study of history of mathematics in the country. However, this chapter would confine itself to the historical context, the nature of the institutions, the curriculum, and aspects of its pedagogy in relation to one particular set of manuscripts, which also constitutes the primary evidence for the reconstruction of the history of mathematics education in the region. The chapter would also briefly outline the nature of the changes that the teaching and learning of mathematics went through during the course of the nineteenth century, against the background of the larger shifts in the colonial policies towards indigenous education.

There are several kinds of manuscripts that could provide a clue to the nature of the mathematical tradition in precolonial India. In case of the Tamil region, there are

(a) The accounts or revenue manuscripts, which use different systems of measures and accounting procedures
(b) The numeracy primers or the *Eṇcuvatis*, which include the *Poṉṉilakkam, Nellilakkam, Eṇcuvati and Kulimāttu*
(c) Mathematical treatises or the *kaṇakkatikāram* manuscripts, which involve a systematic exposition of mathematical engagement, beyond the elementary level.

In this paper, I will mostly talk about the second set of manuscripts – the numeracy primers – *Poṉṉilakkam, Nellilakkam, Eṇcuvati and Kulimāttu*.

The *Poṉṉilakkam* is the most basic numeracy primer. It introduces numbers and notation. In this corpus, we encounter the extensively standardized system of fractions, wherein each fraction below one, can be associated with an area of specific application, either as a measure or as a unit. This extremely complex system of measures represented in terms of numbers in these texts are indicators of the value of precision in a primarily agrarian and mercantile social order where the notion of 'entitlement' was the guiding principle in sharing the produce or the products of labour.

Second, the *Nellilakkam* is the primer that deals with volumetric or cubic measures, where the principle is sequential addition of fixed units, which are then transformed into specific standard units through addition, like *ceviṭu, āḷākku, uḷakku*, etc. Here again there is a strong correlation between the fixed numbers and their practical use in measuring grains, a use that continues even now in several parts of the region.

Third, the *Eṇcuvati* is the quintessential Tamil table book. It usually begins with the multiplication tables of numbers above one. There are two further divisions that are unique to Tamil – multiplication tables of whole numbers by fractions and tables of fractions multiplied by fractions. A typical table will proceed like this: 1×1, 10×1, 2×1, 20×1, ... 9×1, 90×1, 100×1 and at the end of each table there will be a cumulative product which in this case is 595, with a mnemonic verse – 5 jasmine buds bloomed into 90 jasmines shared by five.

Fourth, the *kulimāttu*. This is the table book of squares. This of course, has direct correlations with land measures, which are called *kuḻi, mā* and *vēli* in certain parts and *kāṇi* in certain other parts of the region.

Three different reconstructions are attempted, in order to provide a context for these set of manuscripts.

(a) history of arithmetic practice in the non-institutional context
(b) history of elementary schooling institutions called the *tiṇṇai* schools
(c) history of these texts in relation to these institutions

Through such reconstructions, the chapter will try to bring out the nature of pedagogic engagement with arithmetic knowledge in this period because these manuscripts have significant pedagogical aspects to them.

1. They are dependent on memory as a way of learning rather than as an aid or a skill. This is evident in the extensive system of mnemonics that are not only the constitutive elements of the structure of *Eṇcuvaṭis* but often these mnemonics themselves are algorithms, involving transformations of specific quantities.
2. Such quantities are typically structured as sequences, with addition applied iteratively. These sequences are to be memorized and are marked by the strong presence of mnemonics, which is both a pedagogic mode and an organizing basis.
3. In memory as a mode of learning, memorization, internalization and recollection – will all have to happen in a continuum.
4. But how do you pedagogically accomplish this? By loud recital, reciting after the teacher or the monitor, writing while reciting loudly and through externalized representations as objects.
5. In such a memory mode of learning, writing in particular notations helps the process of visualization that in turn aids associative memory, which is crucial in arithmetic operations. Also, notation and number name together contribute to numeracy.
6. The arithmetic operations are encountered and learned in strong associative relationships and the training aims at associating quantities in relation to each other as one encounters them in the first place. For example, in the number *mā-kāṇi* (1/20 + 1/80), *mā* and *kāṇi* are distinct numbers and units, while at the same time both together as *makāṇi* is also a distinct unit and number. Association in memory becomes central, both in learning numbers as well in problem solving.
7. It is tempting to think of the absence of explicit engagement with transformation and concomitant representations as a mathematical lack. Especially since there are usually no names given to the concepts, nor clear distinctions between processes and products.

Sometimes language provides a clue. For example in certain contexts, for the addition operation, the *ye* sound becomes the mark (*kāni* + *muntiri* = *kāni yē muntiri*) and when multiplication is involved there is no additional sound: *kāni* × *muntiri* = *kāni muntiri*. (*Kāni* and *muntiri* are the fractional quantities 1/80 and 1/320 respectively.) However, one must remember that in the oral tradition, many of the apparent gaps in the written text would have been filled by the teacher or the person who is explaining the text.

8. Language learning and number learning go together since the fixed quantities like *kāni*, *mā* are semantically loaded in the arithmetic context and the learning of such concepts is not divorced from the ordinary day to day use of language of the people at large. This can provide us with important insights into the relationship between the number idea and the development of notation and names.

But what would be the social context that one can reconstruct out of such a mode of learning of mathematics? How can these texts help us to look at the relationship between arithmetic practice, curriculum, pedagogy and the society?

The social ethos of the *tiṇṇai* schools would be reconstructed, to emphasize how its pedagogic strategies were immersed in the cultural context, imbued with shared learning, where creativity in a child is associated with a whole set of agents outside the institution. This is further related to the social status of the household of the child, wherein status is marked by the place of the family in the multiple axes of circulation of common skills of computation. Such skills were crucial in the caste ridden hierarchy of contemporary society, in terms of access to 'entitlements', in other words, products of labour, in the material culture of the locality.

This then compels us to explore the relationship between cognitive universals (as in number learning) and social distribution of ability, strongly associated with questions of access and denial. The chapter will attempt to provide a picture of this complex interaction, by espousing the *tiṇṇai* schools as strongly local institutions, laden with features of access and denial at the same time.

In the remaining section, the chapter will try to outline the several shifts and tensions involving the complex interaction between the techno-economic complex of colonial education and the teaching and learning processes of the indigenous variant. During the course of the nineteenth century, we delineate three broad phases where these tensions would be dealt with. The role that the colonial state apparatus at various layers of its field of action, the missionaries, the European tradition of math teaching, the *tiṇṇai* schools, its teachers, the society in which they functioned all played distinctive roles during this century in attempting to deal with memory and mathematical learning. The way *Eṇcuvati* changed from mnemonic texts to manuals to annexures in the modern textbooks, gaining the new reputation of "bazaar mathematics" along this course will be discussed briefly.

The chapter would conclude by studying one particular case in the early nineteenth century – an attempt to reorganize the *Eṇcuvati*. This, in order to show how a precolonial system of mathematics education, negotiated itself through the onset of colonial educational interventions. In the early nineteenth century, Vēdanāyakam Sāstri educated along with a Maratha prince under the guidance of German missionaries, thought it worthwhile to reorganize the *Eṇcuvati*, and actually did so. While most thought it better not to bother about the details beyond merely changing the Tamil notation into the modern, he reorganized the *Eṇcuvati*, premised on the perceived importance of conceptual understanding of number and its transformations, wherein the processes of transformations of numbers became the guiding principle of reorganization, enabling him to call his work – *Eṇvilakkam*, "number explanation", as opposed to E*n cuvati*- the fixed, static "number text". Even though this attempt, never reached the burgeoning textbook publishing industry through the nineteenth century, and hence never became part of the mainstream, the effort itself showcases the possibilities of a tradition strongly rooted in pedagogical concerns successfully reorganizing itself, through dialogue in the plane of the idea of numbers.

Keywords: history of mathematics education in India, Southern India, history of schools, Tamil

REFERENCES

Eṇcuvati. (1845). Madras: Caturvēta Siddhānta Sabhā. (In Tamil.)
Poṇṇilakkam. (1845). Madras: Caturvēta Siddhānta Sabhā. (In Tamil.)

CHAPTER 75

REPRESENTATIONS OF NUMBERS IN THE INDIAN MATHEMATICAL TRADITION OF COMBINATORIAL PROBLEMS

Raja Sridharan and K. Subramaniam
Tata Institute of Fundamental Research, Mumbai

A rich tradition of combinatorial problems associated with the enumeration of symbol strings and mathematical techniques to solve them has existed in Indian mathematics for over two millennia. These problems have their origin not in a branch of science or technology, but in the arts – in prosody and in music. As far as is known, the first text to deal with such problems is possibly Pingala's *Chandas Shastra*, which deals with enumerating metrical forms of a given length. Pingala's date is uncertain but it is possible that he lived around the time of Panini in the 3rd Century BC. This was a connected (if not continuous) tradition of mathematical work on related problems, that reached a mature mathematical form in the work of Narayana Pandita in the 14th Century CE.

There are several aspects of this tradition that are of potential interest to the mathematics education community. The first is that the mathematics associated with these combinatorial enumeration problems is interesting even from a contemporary perspective, and hence unexpectedly deep. At the same time, large parts of it are accessible without a knowledge of advanced mathematics, and there are several connections with what is learnt at school or in early university education. The second is that numbers in the context of these problems primarily represent not quantity but serial (ordinal) position. That the mathematics associated with such representations can be interesting is a fresh and different perspective that may enrich students' experience of numbers. Finally, the methods used to solve these problems rely on uniquely representing positive integers in a variety of ways, which are vast and interesting extensions of the familiar representations of numbers in base ten or other bases.

An enumeration, called a *prastara*, is a sequence of strings occurring in a fixed order generated by precise rules. The rules are typically recursive – they specify how given a row the next row can be constructed. We begin with some elementary examples of enumerations, which occur in the 14th Century text by Narayana Pandita. A number obviously associated with the prastara is its length or the total number of rows. A prastara corresponding to the number $5 \wedge 2$ is the following table of 25 numbers:

55	34	13	41
45	24	52	31
35	14	42	21
25	53	32	11
15	43	22	
54	33	12	
44	23	51	

In a similar manner one can write down the prastara corresponding to $5 \wedge 3$ starting with 555 and ending with 111.

Historically, the construction of prastaras has its origin in the work of Pingala, as mentioned earlier. Pingala considered prastaras of strings with a fixed syllabic length (i.e. a fixed number of syllables) or *varna vrattas*, where each syllable could be either a short *laghu* '*l*' or a long *guru* '*g*'. Thus the prastara corresponding to a syllable length of 3 consists of the following eight rows:

i) ggg   iv) llg   vii) gll
ii) lgg   v) ggl   viii) lll
iii) glg   vi) lgl

There is a precise rule for enumerating each row, and hence the order of the rows is fixed. Each of the eight rows in the prastara above corresponds to the binary representation of a number between 0 and 7 (written in reverse order). The problem of finding the number of rows with a given syllable length is called *Sankhya*. Two other problems related to a prastara are the *Nashta* and *Uddishta* problems. The *Nashta* problem is to determine the string in the row corresponding to a given row number, while the converse *Uddishta* problem is to determine the row number from the string. It is interesting to observe how the structure of the binary representation is applied to the solution of these two problems for the *varna vrattas*. We also find in Pingala a treatment of the *Lagakriya* problem, that is, to find the number of metres of a given length with a specified number of gurus (or equivalently laghus). This gives rise to the construction of what is now known as the Pascal's triangle (Sridharan, 2005).

Texts dealing with prosody after Pingala, deal with other problems such as constructing prastaras for metres with a fixed number of matras (time units), called *matra vrattas*. These are syllabic strings with a fixed length of time units obtained by taking the short *laghu* 'l' as representing one time unit and *guru* 'g' as representing two time units. The problems associated with *varna vrattas* discussed by Pingala, are now extended to the *matra vrattas* by Virahanka of the 6th or 7th Century CE. The prastara for a metre of three length units for example has 3 rows or strings:

lg, gl, lll

As explained above, the *Sankhya* problem is the problem of determining the number of rows for the prastara of a given metre, that is, the number of possible strings having a given metre length. It is easily checked that the prastara for metres of value 4 contains 5 rows. A recursive relation for the number of rows in the prastara for a metre of length 'n' time units can be found as follows. Group the rows ending in a 'g' and in a 'l' separately. Delete the terminal 'g' in the first group and the terminal 'l' in the second group. It can be shown that from the first, we get the prastara for metres of time length 'n − 2' and from the second we obtain the prastara for metres of length $n - 1$. Thus if $S - n$ is the number of rows in the prastara for the metre of length 'n', then

$$S_n = S_{n-1} + S_{n-2}$$

This is exactly the recursive relation for the so-called Fibonacci numbers. Virahanka's text may well be the first to write down the recurrence relation for the Fibonacci numbers, although it might have been known earlier. Later texts also provide solutions for the problem of Nashta and Uddishta (described above for the varna vrattas) for the matra vrttas. The mathematical rationale underlying the solutions is the fact that any positive integer is either a Fibonacci number or can be expressed uniquely as a sum of non-consecutive Fibonacci numbers (Sridharan, 2006). It is well known that Fibonacci numbers occur widely in many natural contexts. The matra prastara provides a context for grasping the relations among Fibonacci numbers that is accessible.

A treatise dealing with music from the 13th Century CE, the *Sangitaratnakara* of Sarangadeva extends the work described above to musical patterns, both to patterns of musical phrases (swara or musical note combinations) and to rhythm patterns. The former lead to the *tana prastaras* and the latter to the *tana prastaras*. An example of tana prastara considered by Sarangadeva is the enumeration of all phrases containing the swaras S, R, G, M (the first four notes of a 7-note musical scale), where each swara occurs only once. Sarangadeva describes a rule for constructing the rows of the prastara, the number of rows being given by 4!. Next Sarangadeva discusses the Nashta and Uddishta problems for the tala prastaras. The solution to these problems is based on fact that any positive integer m less than on equal to n! can be uniquely represented as follows:

$$m = d_0 0! + d_1 1! + d_2 2! + \ldots + d_{n-1}(n-1)!$$

Where d_i are integers such that $d_0 = 1$ and each d_i lies between 0 and i both inclusive. This is a variant of the general form for the factorial representation of integers (Sridharan, Sridharan & Srinivas, 2010).

Sarangadeva also discusses the tala prastaras or rhythm patterns. The tala patterns are strings of the time units of *druta, laghu, guru* and *pluta*: which have respectively time durations of 1, 2, 4 and 6 units, and are represented as d, l, g and p. A consideration of the Sankhya problem, that is, the number of rows in the prastara for a row with a time length of n, leads to the following recurrence relation:

$$S_n = S_{n-1} + S_{n-2} + S_{n-4} + S_{n-6}$$

The numbers S_n obtained through this relation, which give the number of rows in the prastara for a tala pattern of length n time units, have been called Sarangadeva numbers (Sridharan, Sridharan & Srinivas, 2010). Again, a consideration of the Nashta and Uddishta problems leads to a unique canonical representation of positive integers in terms of the Sarangadeva numbers. Narayana Pandita in his Ganitakaumudi of 1356 CE discusses general recurrence relations of the above type, where $S_n = S_{n-1} + S_{n-2} + \ldots + S_{n-q}$ from a purely mathematical point of view unconnected to applications in prosody or music. Narayana Pandita's work brings this tradition to its culmination.

In the article, we shall provide historical details locating this work in the context of the general tradition of prosody or music. We shall also briefly touch on how aspects of classical music and dance in the contemporary period connect with the older tradition with regard to the use of melodic and rhythm patterns. Drawing from the work of Sridharan and others cited above, we shall present an exposition of the mathematics of the combinatorial problems highlighting the different ways in which numbers are uniquely represented. Finally we shall draw out some of the pedagogical riches that may be found in this tradition.

REFERENCES

Sridharan, R. (2005) Sanskrit Prosody, Pingala Sutras and Binary Arithmetic. In G. G. Emch, R. Sridharan and M. D. Srinivas (Eds.) *Contributions to the History of Indian Mathematics*. Delhi: Hindustan Book Agency.

Sridharan, R. (2006) Pratyayas for Matravrttas and Fibonacci Numbers. *Mathematics Teacher*.

Sridharan, Raja, Sridharan, R. & Srinivas, M. D. (2010) Combinatorial Methods in Indian Music: Pratyayas in Sangitaratnakara of Sarngadeva. In Seshadri, C.S. (Ed.) *Studies in the History of Indian Mathematics*, Delhi: Hindustan Book Agency.